Physical Properties of Biological Membranes and Their Functional Implications

Series of the Centro de Estudios Científicos de Santiago

Series Editor: Claudio Teitelboim

Centro de Estudios Científicos de Santiago
Santiago, Chile
and University of Texas at Austin
Austin, Texas, USA

IONIC CHANNELS IN CELLS AND MODEL SYSTEMS
Edited by Ramon Latorre

PHYSICAL PROPERTIES OF BIOLOGICAL MEMBRANES AND
THEIR FUNCTIONAL IMPLICATIONS
Edited by Cecilia Hidalgo

PRINCIPLES OF STRING THEORY
Lars Brink and Marc Henneaux

QUANTUM MECHANICS OF FUNDAMENTAL SYSTEMS
Edited by Claudio Teitelboim

Physical Properties of Biological Membranes and Their Functional Implications

Edited by
Cecilia Hidalgo
Centro de Estudios Científicos de Santiago
Santiago, Chile

Springer Science+Business Media, LLC

Library of Congress Cataloging in Publication Data

Physical properties of biological membranes and their functional implications / edited
by Cecilia Hidalgo.
   p.   cm. — (Series of the Centro de Estudios Cintíficos de Santiago)
   Series of lectures held in Santiago, Chile on Oct. 1985.
   Includes bibliographies and index.
   ISBN 978-1-4612-8253-2   ISBN 978-1-4613-0935-2 (eBook)
   DOI 10.1007/978-1-4613-0935-2
   1. Membranes (Biology) — Congresses. I. Hidalgo, Cecilia. II. Series.
QH601.P475 1988   87-36111
574.87'5 — dc19   CIP

© 1988 Springer Science+Business Media New York
Originally published by Plenum Press, New York in 1998

Softcover reprint of the hardcover 1st edition 1988

A Division of Plenum Publishing Corporation
233 Spring Street, New York, N.Y. 10013

Contributors

F. J. Barrantes, Instituto de Investigaciones Bioquímicas (INIBIBB), Universidad Nacional del Sur, C.C. 857, 8000 Bahía Blanca, Argentina

Alfred Blume, Institut für physikalische Chemie der Universität Freiburg, D-7800 Freiburg, West Germany

Cecilia Hidalgo, Centro de Estudios Científicos de Santiago, Santiago 9, Chile

Eleonora Kurtenbach, Departamento de Bioquímica, Instituto de Ciencias Biomédicas, Universidade Federal do Rio de Janeiro, 21910 Rio de Janeiro, Brazil

D. Marsh, Max-Planck-Institut für biophysikalische Chemie, Abteilung Spektroskopie, D-3400 Göttingen, Federal Republic of Germany

M. Rojas, Departamento de Biología, Facultad de Ciencias, Universidad de Chile, Santiago, Chile

Mario Suwalsky, Departamento de Química, Universidad de Concepción, Concepción, Chile

Sergio Verjovski-Almeida, Departamento de Bioquímica, Instituto de Ciencias Biomédicas, Universidade Federal do Rio de Janeiro, 21910 Rio de Janeiro, Brazil

F. Zambrano, Departamento de Biología, Facultad de Ciencias, Universidad de Chile, Santiago, Chile

Preface

This book was originated from a series of lectures given in a course on the physical properties of biological membranes and their functional implications. The course was intended to allow students to get acquainted with the physical techniques used to study biological membranes. The experience was valuable and we feel that a detailed description of the procedures used and of various examples of the results obtained allowed many students to become familiar with a theme that is not often part of regular courses on membrane physiology or biophysics.

This book is designed as a tutorial guide for graduate students interested in understanding how physical methods can be utilized to study the properties of biological membranes. It includes first a detailed description of applications of physical techniques—such as X-ray fiber diffraction methods (Chapter 1), ^{2}H and ^{13}C NMR spectroscopy (Chapter 2), and calorimetry (Chapter 3)—in the study of the properties of lipid model membranes. A description of how to measure molecular mobility in membranes (Chapter 4) follows, and the book concludes with three chapters in which biological membranes are the subject of study. Chapter 5 deals with the acetylcholine receptor and its membrane environment; Chapter 6 discusses how fluorescence techniques can be applied in the study of the calcium ATPase of sarcoplasmic reticulum; and Chapter 7 explains how protein-lipid interactions modulate the function of the sodium and proton pumps.

I would like to thank Enrique Jaimovich and Maria Luisa Valdovinos for their help in the organization of the course. My thanks extend as well to the professors and students who made this course possible. The financial help provided by the Tinker Foundation and by Programa de Naciones Unidas para el Desarrollo (PNUD), UNESCO program is gratefully acknowledged.

Cecilia Hidalgo

Santiago, Chile

Contents

Introduction 1

Chapter 1

Structural Studies on Phospholipid Bilayers by X-Ray Fiber Diffraction Methods

Mario Suwalsky

1. Introduction 3
2. X-Ray Fiber Diagrams 4
3. Specimen Preparation................................. 5
4. Photographic Equipment and Technique 6
5. Geometrical Measurement of the Reflections and Information
   Provided 7
6. Intensities of the Reflections 8
7. Electron Density 10
8. Experimental Observations 11
9. Conclusions 17
   References 19

Chapter 2

^2H and ^{13}C NMR Spectroscopy of Lipid Model Membranes

Alfred Blume

1. Introduction   21
2. Principles and Methods   22
3. Applications of ^{13}C and ^2H NMR to Lipid Model Membranes   35
4. Summary..   67
   References   68

Chapter 3

Applications of Calorimetry to Lipid Model Membranes

Alfred Blume

1. Introduction   71
2. Differential Scanning Calorimetry   72
3. Reaction Calorimety   103
4. Conclusions   118
   References   119

Chapter 4

Molecular Mobility in Membranes

D. Marsh

1. Introduction   123
2. Lateral Diffusion   124
3. Rotational Diffusion   133
   References   145

Chapter 5

The Acetylcholine Receptor and its Membrane Environment

F. J. Barrantes

1. Introduction   147
2. Acetylcholine Receptor Primary Structure, cDNA Recombinant
   Techniques, and Acetylcholine Receptor Models   151
3. Acetylcholine Receptor-Mediated Channel Gating: A Single
   Molecule at Work   157
4. The Acetylcholine Receptor in the Lipid Bilayer   159
5. Dynamics of Acetylcholine Receptor and Lipids in the Membrane   168
   References   171

Chapter 6

Fluorescence Spectroscopy in the Study of Sarcoplasmic Reticulum Calcium-ATPase

Sergio Verjovski-Almeida and Eleonora Kurtenbach

1. Introduction 177
2. Structure and Conformation 178
3. Macromolecular Association 196
4. Protein–Protein and Protein–Lipid Interactions 204
   References 207

Chapter 7

Lipid–Protein Interactions in the Function of the Na^+ and H^+ Pumps: Role of Sulfatide

F. Zambrano and M. Rojas

1. Introduction 211
2. The Sodium Pump 212
3. Lipid Requirements of the Sodium Pump: Role of Sulfatides .. 213
4. Role of Sulfatides in the Proton Pump 228
5. Conclusions 231
   References 231

Index 233

Introduction

Cecilia Hidalgo

Biological membranes have complex lipid compositions, with phospholipids and cholesterol as their main components. More than 100 different lipids have been found in biological membranes. The reasons for the presence of such a large variety are not well understood, and only in a few cases have specific functions been attributed to membrane lipids.

In order to understand the behavior of lipids in biological membranes, many studies have been carried out with model membranes composed either of a single lipid class or of simple mixtures with two to three components. Several physical techniques have been used to study these model membranes, among them X-ray diffraction; differential scanning calorimetry; and fluorescence, NMR, and EPR techniques. A detailed picture of how lipids are organized in membranes has emerged from these studies. Thus, the packing arrangement of phospholipids and cholesterol in bilayers is now better understood, and such subjects as the differences in the physical properties of unsaturated phospholipids and saturated compounds, and how cholesterol affects these properties, are becoming increasingly clear.

The knowledge gained in studies on model lipid membranes has been instrumental in understanding the properties of biological membranes. Many studies using a variety of physical techniques have been carried out with either intact biological membranes or the lipids extracted from them.

CECILIA HIDALGO • Centro de Estudios Científicos de Santiago, Santiago 9, Chile.

From these studies it is now known that membrane proteins affect the mobility of the lipids in their immediate vicinity, and that some membrane enzymes have specific lipid requirements for function. Furthermore, the physical state of membrane lipids has a profound influence on the properties of membrane proteins. As an example, studies using membrane enzymes reconstituted with well defined synthetic phospholipids have shown that the physical state of the lipids modulates enzyme activity, so that lipids in the gel phase in general cause a drastic inhibition of enzyme activity. These findings indicate that a fluid lipid phase, which allows both lipid and protein mobility in the phase of the membrane, is needed to allow membrane proteins to function under optimal conditions. Using physical techniques it has been established that, at physiological temperatures, biological membranes do indeed have a fluid lipid phase. It is possible that the diversity of lipid molecules found in biological membranes ensures a fluid lipid phase under the variety of conditions that cells encounter during their life cycle.

As our knowledge of the physical properties of biological membranes increases, we shall be better able to understand how these properties affect, for instance, lipid–protein interactions, which, in turn, modulate the function of membrane proteins.

Chapter 1

Structural Studies on Phospholipid Bilayers by X-Ray Fiber Diffraction Methods

Mario Suwalsky

1. INTRODUCTION

Biological membranes are mainly composed of lipids and proteins, with phospholipids playing a key role in their structure and function. These are zwitterionic molecules of amphipathic nature, which generally tend to form bilayers somewhat similar to those present in the cellular membranes. Because of their chemical complexity and the nonperiodical arrangement of the constituent molecules, natural membranes generally do not form either crystals or well-oriented specimens suitable for X-ray structure analysis. Phospholipid bilayers constitute, therefore, useful models for obtaining information about the assembly and function of biological membranes.

The red cell membrane has been the object of structural investigations for many years. It consists of approximately equal weights of lipids and proteins. It is well recognized that the phospholipids are organized as an asymmetric bilayer in which diacylphosphatidylethanolamines and phosphatidylserine are preferentially localized at the inner half of the membrane bilayer while diacylphosphatidylcholines and sphingomyelin are principally in the outer leaflet. The functional significance of bilayer asymmetry and its maintenance is not yet fully understood (Marchesi et al., 1976).

MARIO SUWALSKY • Departmento de Química, Universidad de Concepción, Concepción, Chile.

3

In our laboratory we have been studying the molecular conformation and packing arrangement of different phospholipids related to red cell membranes by X-ray fiber diffraction and electron microscopy (Suwalsky and Tapia, 1981; Suwalsky and Knight, 1982; Suwalsky *et al.*, 1984; Suwalsky and Tapia, 1985; Suwalsky and Duk, 1987). Specimens prepared in the form of oriented films, fibers, and powders have been analyzed under a wide range of hydration at room temperature.

It is the purpose of this chapter to describe the methods and techniques employed to study the bilayer structure of various phospholipids by X-ray fiber diffraction and discuss the observed results.

2. X-RAY FIBER DIAGRAMS

This kind of X-ray diffraction pattern is generally produced by specimens formed by long molecules that do not form single crystals such as DNA, polysaccharides, fibrous proteins, synthetic polymers, etc. However, they might tend to align more or less parallel to each other forming samples with a certain degree of order. These specimens may present an order almost crystalline in certain microscopic domains known as "crystallites." If they are randomly oriented, series of concentric circles known as powder patterns will be produced after subjecting these specimens to X-ray diffraction.

Through certain special techniques it is possible to orient the "crystallites" along one or more preferential directions. This is usually attained by pulling fibers or growing oriented films from solutions. In the first case the longest axes of the molecules lie more or less parallel to the fiber axis, but otherwise the "crystallites" present rotational disorder. The effect of irradiating a static fiber with monochromatic X rays is equivalent to doing it to a single crystal rotating around an axis perpendicular to the incident beam.

Another way of orienting specimens is the growing of oriented films. In this case the "crystallites" may lie with the long molecular axes in the plane of the film without a preferential direction (unidimensional order). The "crystallites" might also lie not only in the plane of the film but also with their long axes parallel to each other (bidimensional order). These specimens are usually oriented edge on (parallel) to the incoming X-ray beam to generate X-ray fiber diagrams.

This kind of diagram presents significant differences with respect to those obtained from single crystals. Whereas the latter show many reflections (generally between 30 and 50 per atom), which look like discrete points, the fiber diagrams generally exhibit few reflections, which take the form of broad and diffuse arcs whose lengths are a function of the degree of orientation of the "crystallites."

The reflections in the fiber diagrams are distributed in layer lines. Those located on the equator (0 layer line) are known as equatorial reflections, while those on the meridian are called meridional reflections (see Fig. 1).

The analysis of the diagrams is based on the measurement of the position and intensity of each reflection. The former provides information that leads to the calculation of the observed interplanar spacings (d_o), type and dimensions of the unit cells, and their space groups. With the aid of chemical analysis, density measurements, model building, and any other available data it is possible to calculate the number of molecules in the unit cells and obtain preliminary information about their conformation and modes of packing. On the other hand, the intensities of the reflections allow us to calculate and plot electron density maps or profiles which provide the basic information to determine as precisely as possible the molecular structure involved.

3. SPECIMEN PREPARATION

Synthetic L-α-dilauroyl-(DLPE), dimyristoyl-(DMPE), dipalmitoyl-(DPPE) phosphatidylethanolamines and L-α-dimiristoylphosphatidylcholine (DMPC) from Sigma and Calbiochem were used. Oriented specimens in the form of thin films were prepared from $CHCl_3$ or $CHCl_3 : CH_3OH$ 1 : 1 solutions by a procedure developed in this laboratory. It consists in the careful and slow evaporation of the solvent in closed systems until thin films are formed on their surfaces. They are then collected with the aid of Pt or Nicrom fine rings and left to dry. Then they are cut with a surgical

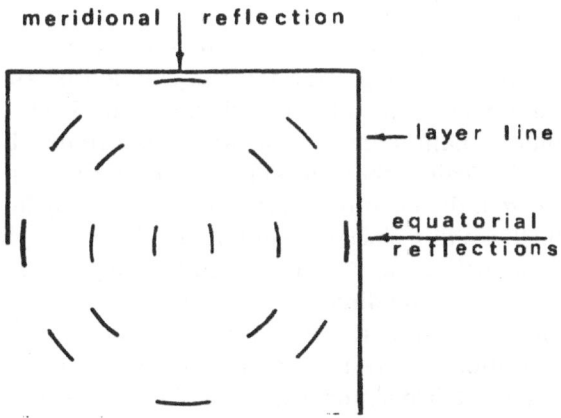

Figure 1. Schematic representation of a fiber diagram.

knife into small rectangular pieces of about 1 mm^2 and fixed with glue to the end of thin glass fibers. The specimens thus prepared are placed in small containers connected to saturated solutions which produce constant relative humidities (r.h.) of 33%, 47%, 86%, and 92% at room temperature (about 18°C). With P$_2$O$_5$ 0% r.h. is obtained. The samples are kept at each r.h. for several weeks until equilibrium is reached. This is determined by gravimetry in parallel samples, which are also used to calculate the corresponding degree of hydration and density at each r.h., the latter being determined by flotation. Completely dried specimens are obtained by heating at vacuum for 4 hr at 90°C over P$_2$O$_5$.

Samples are also exposed to the highest possible hydration by immersing them in sealed 1.5-mm-diameter thin-walled low X-ray absorbing glass capillaries filled with distilled water.

4. PHOTOGRAPHIC EQUIPMENT AND TECHNIQUE

Oriented diffraction patterns are obtained using flat-plate cameras with nickel-filtered Cu K_α radiation from a Philips fine focus tube. The collimation is provided by a Pyrex glass capillary 4.5 cm long with inner diameter 0.25 mm, mounted in a 5.0-cm-long brass tube with pinholed lead-disk guards at both ends of the collimator. A thin piece of Mylar is placed between the collimator and the external guard to avoid leakage of the gas that is used to fill the camera or clogging by water or crystals that could be carried over by the gas. Diffraction from the Mylar and capillary edges is intercepted by the guard. This system is mounted in a holder that allows for the proper alignment of the collimator. A cylinder that could fit in the collimator and has X-ray fluorescent powder and a piece of lead glass at the other end is used to align it.

The specimens are photographed in a humidity-controlled cell in equilibrium with the corresponding saturated salt solution of known r.h. (Fig. 2). The collimator fits directly into the cell and a thin sheet of Mylar serves as an exit window. Diffraction by the Mylar is avoided by placing the beam-stop directly against the exit window. The beam-stop consists of a 1-cm-long and 1-mm-diameter brass tube filled with X-ray fluorescent powder. It is attached to a thin sheet of celluloid fixed onto the cassette holding the X-ray film in such a way that it can quite easily be set in the proper position with respect to the direct X-ray beam.

The specimens are mounted on holders that allow them to be oriented at any desired position with respect to the X-ray beam; most generally they are aligned edge on. Nitrogen or helium bubbled through a saturated salt solution that produces a constant r.h. is circulated through the cell, which also contains open vessels with the same salt solution. Specimens-to-film

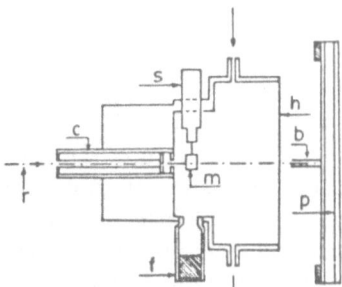

Figure 2. Schematic drawing of the humidity-controlled flat-plate camera. c, collimator; r, X-rays; m, sample; s, sample support; f, saturated salt solution; b, beam-stop; p, photographic film; h, sheet of Mylar. Arrows indicate gas flow.

distance are generally about 60 mm and are calibrated by sprinkling a little calcite powder on the specimens.

5. GEOMETRICAL MEASUREMENT OF THE REFLECTIONS AND INFORMATION PROVIDED

The horizontal distance ($2x$, in mm) between the centers of the two corresponding arcs of each equatorial reflection is measured as accurately as possible. The observed interplanar spacings, d_o, are obtained through the equation

$$d_o = \frac{\lambda}{2 \sin[\frac{1}{2} \arctan(x/D)]}$$

where λ is the long-wave of the X-ray beam (for Cu K_α is about 1.542 Å), and D is the specimen-to-film distance.

If the oriented film of a phospholipid has been exposed to the X rays in a vertical position and parallel to the beam, the equatorial reflections might provide information about the length of one or two of the unit cell basal axes. The longest one will be related to the bilayer width, which generally is of the order of 50–60 Å. In the case of helical structures additional information related to the packing and diameter of the helices can also be gathered.

The interplanar spacings of the nonequatorial reflections can be calculated from the measurements of the horizontal ($2x$) and vertical ($2y$) distances between every reflection in each quadrant through the equation

$$d_o = \frac{\lambda}{(\zeta^2 + \xi^2)^{1/2}}$$

where the reciprocal cell coordinates ζ and ξ can be obtained from

$$\zeta = \frac{y}{(D^2 + x^2 + y^2)^{1/2}}$$

and

$$\xi = \left| 2 - \zeta^2 - \frac{2D(1-\zeta^2)^{1/2}}{(D^2 + x^2)^{1/2}} \right|^{1/2}$$

The analysis of the nonequatorial reflections completes the information about the unit cell dimensions. Moreover, by looking for all the systematically absent reflections the space group can be determined. In the case of helical structures, nonequatorial reflections can help to find the axial translation of the monomers and the pitch of the helix.

6. INTENSITIES OF THE REFLECTIONS

The precise atomic coordinates can be found only through the analysis of the intensities of the reflections.

The Bragg equation $2d \sin \theta = n\lambda$, where θ is the diffraction angle and n, an integer, is the order of the reflection, describes the conditions under which X rays are diffracted by the crystal planes. If every atom were located exactly on these planes, all the diffracted beams would be of the same maximum intensity. That this is not the case can be easily appreciated just by looking at any X-ray diagram where the intensities abruptly and discontinuously change from one reflection to another.

This is exactly what is to be expected if a number of atoms are not precisely located on the crystal planes but distributed in the volume of the unit cell. It is possible to have some crystal planes that include or pass near many atoms producing high-intensity reflections. However, the waves coming from the atoms located between the planes will be out of phase with respect to those produced in the planes. As a result the diffracted beam will be weakened or canceled out. In other words, the particular distribution of the atoms in the volume of the unit cell is the main factor that determines the intensities of diffracted beams.

The fundamental problem, however, is to determine the position of the atoms from the observed intensities, for which there is no easy direct solution.

If there is a group of atoms P_0, P_1, $P_2, \ldots, P_n$ in the unit cell (Figs. 3a and b), we can arbitrarily choose one (P_0) as the origin of a given plane. If P_0 diffracts, P_0'—which is the equivalent atom in the next plane—will diffract too. It can be assumed that these two atoms define the a axis and that the path difference between the beams diffracted by P_0 and P_0' is equal to $2a \sin \theta = 1\lambda$ for the first-order reflection, or more generally, $2a \sin \theta = h\lambda$ for each reflection of order h. The phase difference of the waves coming from the atoms P_0 and P_0' is $2\pi h$ and the net result will be a beam with twice the original intensity.

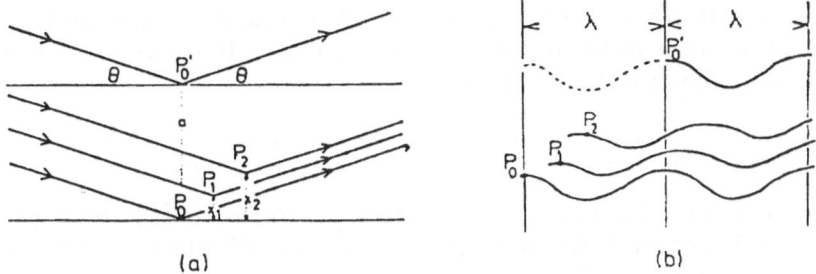

Figure 3. (a, b) Representation of the contribution of atoms P_0, P_1, P_2, ..., to the structure amplitude F.

In the case of the atom P_1, located at a distance x_1 from the plane, the path difference of the diffracted wave with respect to that coming from P_0 is

$$2x_1 \sin \theta = \frac{h\lambda x_1}{a}$$

and the difference of phase in radians will be

$$\frac{2\pi hx_1}{a}$$

This is equally valid for the rest of the atoms.

Each wave is characterized by an amplitude $f_0, f_1, f_2, \ldots, f_n$, which depends on the dispersion power of each atom and of the angle of the diffracted beam; it is known as the atomic scattering factor of each atom and depends on its atomic number. The structure amplitude $|F|$ is the amplitude that results from the summation of all the waves dispersed by the contents of the unit cell. Figure 4 shows the vectorial summation of the wave amplitudes.

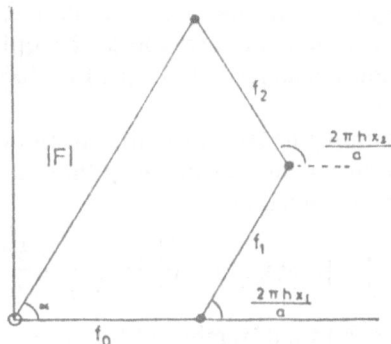

Figure 4. Vector composition of the structure amplitude F.

Analytically, $F(h\,0\,0) = \sum_j f_j e^{2\pi i h x_j/a}$. For a general plane $(h\,k\,l)$ the change of phase from the origin to a point $x_1\,y_1\,z_1$ is $2\pi(hx_1/a + hy_1/b + lz_1/c)$ and the resulting vector is

$$F(h\,k\,l) = \sum_j f_j \exp\left[2\pi i\left(\frac{hx_j}{a} + \frac{ky_j}{b} + \frac{lz_j}{c}\right)\right]$$

This complex equation, characterized by an amplitude $|F|$ and a phase angle α, is known as the Structure Factor. It can be calculated by means of the equation

$$|F(h\,k\,l)| = (A^2 + B^2)^{1/2}$$

where

$$A = \sum_j f_j \cos 2\pi\left(\frac{hx_j}{a} + \frac{ky_j}{b} + \frac{lz_j}{c}\right)$$

$$B = \sum_j f_j \sin 2\pi\left(\frac{hx_j}{a} + \frac{ky_j}{b} + \frac{lz_j}{c}\right)$$

These summations include all the atoms in the unit cell. If the structure is centrosymmetric and the center of symmetry is at the origin of the unit cell all vectors with the phase angle $2\pi x_j/a$ are accompanied by other vectors of phase angle $-2\pi x_j/a$. The resulting vector lies on the abscissa and its phase angle is limited to 0 or π. Therefore, $B = 0$ in the centrosymmetric structures and $F(h\,k\,l)$ is the sum of twice the cosine and the summation is taken for half the total number of atoms in the unit cell.

7. ELECTRON DENSITY

The electron density, $\rho(x\,y\,z)$, is defined as the number of electrons in a given point x, y, z of the unit cell. This density is expressed in such a way that $\rho(x\,y\,z)\,dx\,dy\,dz$ corresponds to the number of electrons in the volume $dx\,dy\,dz$. For the general case where the unit cell axes are non-orthogonal, the function is changed to $\rho(x\,y\,z)\,(V/abc)\,dx\,dy\,dz$, where V is the unit cell volume.

Assuming a continuous density of matter in the unit cell instead of a discrete number of atoms j located in the points x_j, y_j, z_j, the structure factor $F(h\,k\,l)$ can be expressed as

$$F(h\,k\,l) = \frac{V}{abc}\int_0^a \int_0^b \int_0^c \rho(x\,y\,z)\exp\left[2\pi i\left(\frac{hx}{a} + \frac{ky}{b} + \frac{lz}{c}\right) dx\,dy\,dz\right]$$

changing the sum over the total number of atoms by a triple integral over the whole volume of the unit cell. In this case the phase difference is expressed as

$$2\pi\left(\frac{hx}{a} + \frac{ky}{b} + \frac{lz}{c}\right)$$

To solve a structure we must then evaluate its electron density, which can be expressed by the inverse Fourier transform of $F(h\,k\,l)$:

$$\rho(x\,y\,z) = \sum_{-\infty}^{\infty}\sum_{-\infty}^{\infty}\sum_{-\infty}^{\infty} \frac{|F(h\,k\,l)|}{V} \exp\left[-2\pi i\left(\frac{hx}{a} + \frac{ky}{b} + \frac{lz}{c}\right)\right]$$

This equation can also be expressed as

$$\rho(x\,y\,z) = \sum_{-\infty}^{\infty}\sum_{-\infty}^{\infty}\sum_{-\infty}^{\infty} \frac{|F(h\,k\,l)|}{V} \cos\left[2\pi\frac{hx}{a} + 2\pi\frac{ky}{b} + 2\pi\frac{lz}{c} - \alpha(h\,k\,l)\right]$$

The simplest expression of the electron density is that confined to one dimension (z axis), where the equation becomes

$$\rho(z) = \sum_{\infty}^{\infty} \frac{|F(0\,0\,l)|}{d\,0\,0\,1} \cos\left[2\pi\frac{lz}{c} - \alpha(0\,0\,l)\right]$$

and originates what is called an electron density profile, which corresponds to a projection of the number of electrons along one of the crystal axes.

In the centrosymmetrical structures, this equation takes the form

$$\rho(z) = \frac{2}{d_{001}} \sum_{l=0}^{\infty} \psi(l)F(l) \cos 2\pi\frac{lz}{c}$$

where $\psi(l)$, the phase angle, is $+$ or $-$.

8. EXPERIMENTAL OBSERVATIONS

Oriented films of the phospholipids under study have been subject to X-ray fiber diffraction analysis in a wide range of hydration at room temperature (about $18°C \pm 1$). The most crystalline patterns are observed at 33% and 47% r.h. showing between 50 and 100 nearly crystalline and well-oriented reflections. Figure 5 shows one of such a pattern.

The crystallographic and physicochemical parameters of each phospholipid at different relative humidities, including the crystal system, space group, unit cell dimensions, number of phospholipid molecules in the cell (Z), S, the molecular area at the bilayer surface [calculated from the expression $S = ab \sin(\gamma/Z)$], number of molecules of water per molecule of phospholipid, and both theoretical (dt) and experimental (dx) densities, are presented in Tables I–IV.

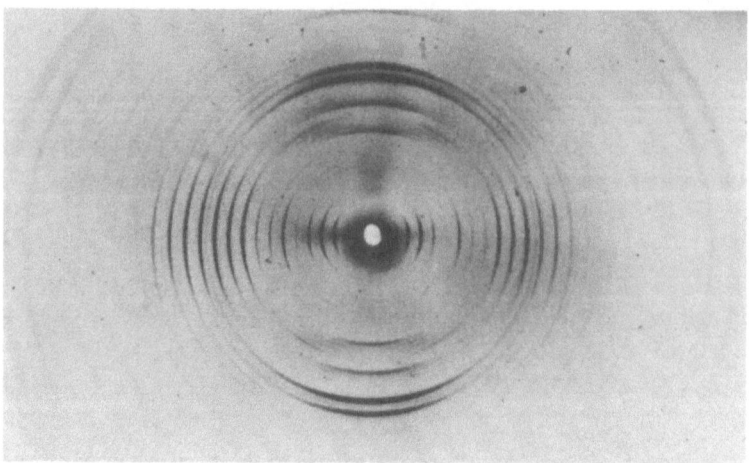

Figure 5. X-ray fiber diagram of an oriented film of DMPE, edge on, at 47% relative humidity, 19°C.

When oriented films of the four phospholipids under study are introduced into excess of liquid water, DMPC and the diacylphosphatidylethanolamines show a remarkable difference. It is observed that the c axis (bilayer repeat) of DMPC increases from about 55 Å at 92% r.h. to 63.5 Å after only 1 hr of exposure to water. This value increases to 78 Å after 5 hr and to 87 Å after 21 hr. All these changes are accompanied by a sharp decrease in the number of reflections. After 60 days only a 4.2-Å unoriented

Table I. DLPE: Unit Cell and Physicochemical Parameters as a Function of Hydration[a]

	Percent relative humidity		
	0	47	92
a (Å)	8.00	7.99	8.00
b (Å)	9.65	9.80	9.85
$\frac{1}{2}c$ (Å)	46.17	46.05	46.32
γ	90°	90°	90°
$\frac{1}{2}V$ (Å³)	3565	3606	3878
$\frac{1}{2}Z$	4	4	4
S (Å²)	38.6	39.2	39.4
$nH_2O/DLPE$	0	1.0	1.7
dt (g/cm³)	1.08	1.10	1.11
dx (g/cm³)	1.06	1.07	1.08

[a] Crystal system: Monoclinic pseudo-C-centered; space group: P_2.

Table II. DMPE: Unit Cell and Physicochemical Parameters as a Function of Hydration[a]

	Percent relative humidity			
	0	33	47	86
a (Å)	7.45	7.50	7.53	7.56
b (Å)	9.70	9.80	9.82	9.82
c (Å)	50.25	50.27	50.75	51.10
γ	90°	90°	90°	90°
V (Å^3)	3631	3695	3753	3794
Z	4	4	4	4
S (Å^2)	36.1	36.7	37.0	37.1
$n\,H_2O/DMPE$	0	0.4	0.8	1.3
dt (g/cm^3)	1.17	1.14	1.15	1.13
dx (g/cm^3)	1.10	1.10	1.12	1.12

[a] Crystal system: Monoclinic pseudo-C-centered; space group: P_2.

ring is observed. On the other hand, the c axes of the acylphosphatidyl-ethanolamines only increase by about 1 Å after 30 days of exposure to water and still show more than a dozen sharp reflections.

The structure of the phospholipids is determined as explained previously. The intensities of the equatorial reflections of the most crystalline patterns (I_o) are measured with the aid of a microdensitometer. The structure factors $|F_o|$ of most of the equatorial reflections are corrected by the Lorenz polarization factor $L_p = (1 + \cos^2 2\theta)/\sin 2\theta$ and the temperature factor

Table III. DPPE: Unit Cell and Physicochemical Parameters as a Function of Hydration[a]

	Percent relative humidity			
	0	33	43	86
a (Å)	8.05	8.23	8.32	8.50
b (Å)	9.92	10.18	10.24	10.24
c (Å)	55.50	55.75	55.98	56.00
γ	90°	90°	90°	90°
V (Å^3)	4432	4671	4769	4874
Z	4	4	4	4
S (Å^2)	39.9	41.9	42.6	43.5
$n\,H_2O/DPPE$	0	0.6	2.7	2.9
dt (g/cm^3)	1.04	1.00	1.04	1.02
dx (g/cm^3)	1.02	1.02	1.02	1.00

[a] Crystal system: Monoclinic pseudo-C-centered; space group: P_2.

Table IV. DMPC: Unit Cell and Physicochemical Parameters as a Function of Hydration[a]

	Percent relative humidity			
	33	56	76	92
a (Å)	9.34	9.34	9.27	9.40
b (Å)	9.81	9.65	9.85	10.03
c (Å)	54.60	54.38	54.97	54.84
γ	96.0°	96.4°	96.4°	96.0°
V (Å^3)	4975	4871	4988	5142
Z	44	4	4	4
S (Å^2)	45.6	44.8	45.4	46.7
$n H_2O/DMPC$	3.3	3.4	5.7	7.9
dt (g/cm^3)	0.99	1.01	1.04	1.06
dx (g/cm^3)	1.03	1.03	1.01	1.01

[a] Crystal system: Monoclinic pseudo-C-centered; space group: P_2.

$T = \exp[-B(\sin \theta/\lambda)^2]$ with B, the temperature constant, set equal to 3.0. The phase angles of the equatorial reflections are determined by trial and error on the basis of a graphical procedure described in detail elsewhere (Caspar and Kirschner, 1971). Since the bilayers are centrosymmetric the phase angles are restricted to 0 or π, i.e., each structure factor has either a + or − sign. The observed structure factors $|F_o|$ are plotted against $1/dc$ (the reciprocal of the calculated interplanar spacings). The resulting curve is shown in Fig. 6. The change of sign of the modulus occurs at $|F_o| = 0$; the only choice is to decide whether to start with a + or − sign, which is assigned by trial and error. The plot of its electron density profile $\rho(z)$ is shown in Fig. 7a.

This electron density profile shows the following characteristics:

1. The highest peak is located about 23 Å from the origin; it should correspond to the polar head, which, having P, O, and N atoms, is

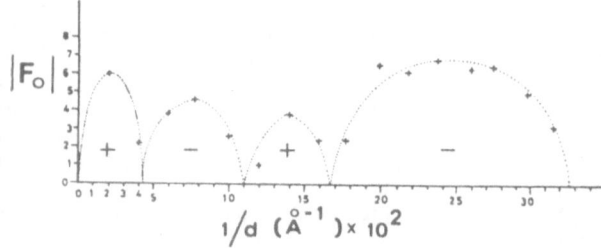

Figure 6. Plot of observed structure factors F_o versus the reciprocal of equatorial interplanar spacings of DMPE.

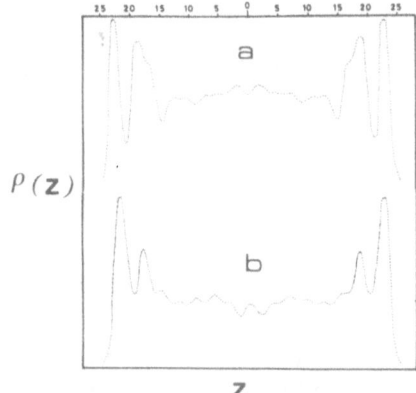

$P(z)$

Figure 7. Electron density profiles for DMPE at 47% relative humidity. (a) Based on observed structure factors F_o; (b) based on calculated structure factors F_c.

z

  the chemical group of the phospholipid with the highest electron density.

2. There is another peak, somewhat lower and broader, at about 19 Å, which might be assigned to the two carboxyl groups.
3. There is an almost flat plateau extending for about 15 Å that could most certainly be attributed to the long and extended hydrocarbon chains.

  With this information and data available plausible molecular models are built from skeletal and CPK components. This model building serves to narrow down the possibilities to a small number of structures, from which preliminary atomic coordinates are obtained. Theoretical models are then generated by means of a computer program by systematically varying the torsion angles of the three phospholipid chains. In each case the atomic coordinates, structure factors $F_c(l)$, and electron densities are calculated. The corresponding electron density profiles are compared with that obtained from the observed structure factors $F_o(l)$ (Fig. 7a). After the introduction of a molecule of water in the hydrophilic space formed by the polar and glycerol groups the refinement process is stopped given the similarity of this profile (Fig. 7b) with that shown in Fig. 7a.

  From the comparison of both electron density profiles it is concluded that the theoretical model should be very close to the molecular conformation of this phospholipid (DMPE) in the oriented film form at 47% r.h. Figure 8 shows its skeleton model, which presents the following characteristics:

1. The polar head and the two hydrocarbon chains are coplanar to the xz plane.
2. The terminal diacyclphosphatidylethanolamine group lies perpendicularly with respect to the hydrocarbon chains.

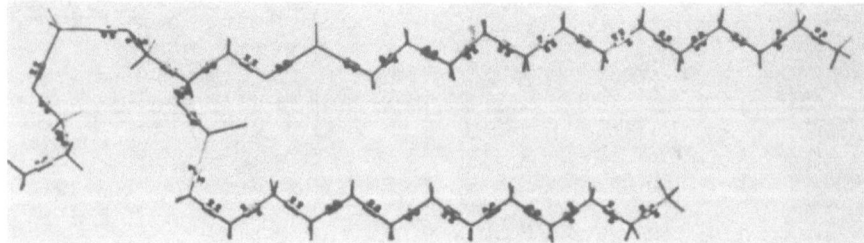

Figure 8. Skeleton model of a DMPE molecule.

3. While one of the hydrocarbon chains is completely extended, the other one looks shorter as it makes an almost right-angle turn in its initial part.
4. One of the hydrocarbon chains is somewhat rotated along its longest axis with respect to the other hydrocarbon chain.

The packing arrangement of the phospholipids are studied with the aid of space filling models. For the sake of comparison we present in Figs. 9 and 10 the molecular packing of diacylphosphatidylethanolamine (DMPE) and that of lecithin (DMPC), respectively; the phospholipids only differ in their terminal amino group. In the case of DMPE the molecules pack very tightly forming a typical bilayer structure 50.75 Å thick. Contacts between monolayers occur via the terminal methyl groups as well as through electrostatic interactions between the charged PO_4^- and NH_3^+ groups. There is also a considerable number of hydrophobic interactions between neighboring hydrocarbon chains. This arrangement leaves a space barely enough to accommodate just one molecule of water per DMPE, in agreement with the experimental data. The three-dimensional model is built by placing one bilayer on top of the other in the y direction. The length of the b axis, 9.82 Å, is about twice the expected value. This is due to the fact that the sheets are piled up in the y direction, or alternatively shifted along x by about $a/2$. The only elements of symmetry present in this arrangement are

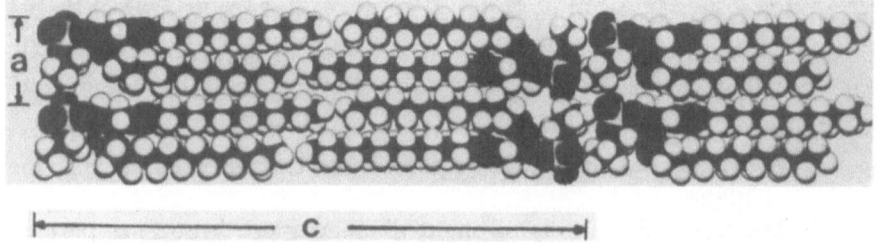

Figure 9. Two-dimensional molecular model of a DMPE bilayer.

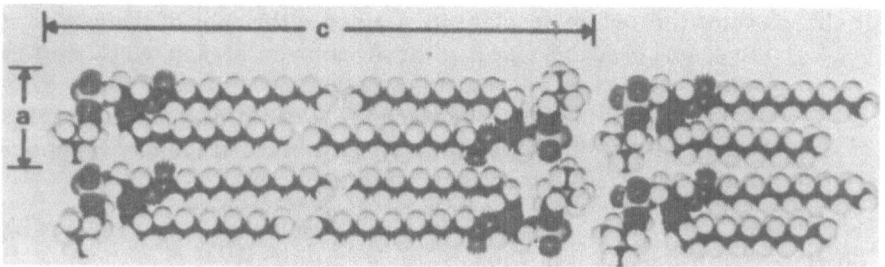

Figure 10.   Two-dimensional molecular model of a DMPC bilayer.

twofold rotation axes normal to the xz plane. The space group is, therefore, P_2; the unit cell is monoclinic with orthogonal axes, pseudocentered in the C phase.

The structural features described for DMPE are very similar to those reported for the oriented films of dilauroyl- (DLPE) and dipalmitoylphosphatidylethanolamine (DPPE). They mostly differ from each other by about 5 Å in their bilayer repeat distance, which approximately corresponds to the length of four methylene groups in the hydrocarbon chains (see Tables I–III).

On the other hand, the molecular conformation of the lecithin DMPC in the oriented-film form does not differ too much from the diacylphosphatidylethanolamines (Suwalsky and Tapia, 1985). However, its packing arrangement is less compact than the latter. (Compare Figs 9a and b and Tables I–IV.)

9. CONCLUSIONS

The procedures to solve the structure of phospholipid bilayers by X-ray fiber diffraction methods have been described in some detail. The structures of four phospholipids have also been described: three diacylphosphatidylethanolamines (DLPE, DMPE, and DPPE), which differ from each other only in two methylene groups in their hydrocarbon chains, and a lecithin (DMPC).

An analysis of the results permits us to conclude that the four phospholipids show several common structural features, including the following:

1. Their hydrocarbon chains are mostly parallel and fully extended, packed as closely as possible.
2. The polar groups are coplanar and lie perpendicularly with respect to the hydrocarbon chains.

3. The molecules stack laterally to form monolayers. Bilayers are formed through molecular interactions at the ends of the hydrocarbon chains in such a way that the monolayers are related by twofold rotation axes perpendicular to the planes.
4. Hydrophobic interactions between the hydrocarbon chains and electrostatic attractions between opposite charges of neighboring polar groups stabilize the bilayer structure.

At the same time the four phospholipids present some interesting differences, such as the following:

1. The bilayer repeat distances of the acylphosphatidylethanolamines DLPE, DMPE, and DPPE differ in that order by about 5 Å. This is not surprising as this distance approximately corresponds to the length of four methylene groups, by which their hydrocarbon chains differ.
2. The packing is closer in the acylphosphatidylethanolamines than in DMPC, an effect that might be due to the smaller volume of their terminal amino groups.
3. DMPC presents higher degrees of hydration than the other phospholipids at about the same relative humidities. On the other hand, an excess of hydration does not seem to affect significantly the packing arrangement of the acylphosphatidylethanolamines as compared to DMPC.

All these similarities and differences might be related to the functional role played by the phospholipid bilayers in the cellular membranes. In this sense it could be that lecithins, with their bulkier polar groups, are located in the external part of the red blood cell membranes and the diacylphosphatidylethanolamines in the cytoplasmic side just to contribute to the proper curvature of the cell. On the other hand, the higher stability of the structures of diacylphosphatidylethanolamines in the presence of excess of liquid water might be necessary to have stable interactions between the cell membrane and the supporting cystoskeleton.

Presently, the structural studies on phospholipid bilayers have been extended to include their interactions with therapeutic drugs and chemicals of biological interest. Compounds such as DDT, antibiotics, and radioprotectors are made to interact with the lipids in order to understand their effects on the bilayers of cell membranes (Suwalsky *et al.*, 1985).

ACKNOWLEDGMENTS. Support by research grants from the Research Council of the University of Concepción (2.15.40 and 20.13.27) and the Volkswagen Foundation (11-1744) is gratefully acknowledged.

REFERENCES

Caspar, D. L. D., and Kirschner, D. A., 1971 *Nature* **231**:46-52

Marchesi, V. T., Furthmayr, F., and Tomita, M., 1976, *Annu. Rev. Biochem.* **45**:667-698.

Suwalsky, M., and Duk, L., 1987, *Makromol. Chem.* **188**:599-606.

Suwalsky, M., and Knight, E., 1982, *Z. Naturforsch.* **37c**:1156-1160.

Suwalsky, M., and Tapia, J., 1981, *Z. Naturforsch.* **36c**:875-879.

Suwalsky, M., and Tapia, J., 1985, *Bol. Soc. Chil. Quím.* **30**:27-35.

Suwalsky, M., and Seguel, C. G., and Neira, F., 1984, *Z. Naturforsch.* **39c**:141-146.

Suwalsky, M., Bugueño, N., Tapia, J., and Neira, F., 1985, *Z. Naturforsch.* **40c**:566-570.

Chapter 2

²H and ¹³C NMR Spectroscopy of Lipid Model Membranes

Alfred Blume

1. INTRODUCTION

The elucidation of structure and dynamics of biological and model membranes has been an area of intensive research over the last years. As biological membranes are very complex and also difficult to isolate in sufficient quantities needed for certain types of physicochemical techniques, many experiments have been performed with model membranes of defined chemical composition. While calorimetry provides information on the thermodynamic properties of the system and X-ray techniques can elucidate the structural aspects of the bilayers, the dynamics of the lipid molecules can be elucidated by spectroscopic techniques. In particular, NMR spectroscopy is a powerful technique because it is an intrinsic method and does not require the introduction of a potentially perturbing label or probe as is the case for ESR and fluorescence spectroscopy. Particularly ²H and ¹³C NMR spectroscopy are well suited for several reasons:

1. The ²H or ¹³C nuclei can be introduced into the molecules under study via chemical synthesis, so that either only one resonance is observed (in the case of deuterium) or the spectrum is dominated by the line of the ¹³C nucleus in the enriched site.

ALFRED BLUME • Institut für physikalische Chemie der Universität Freiburg, D-7800 Freiburg, West Germany.

2. The chemical shift anisotropy of ^{13}C provides the possibility of studying motions in the 500–5000-Hz range.
3. The quadrupole interaction of the ^2H nucleus is in the range of 2×10^5 Hz and deuterium spectra are therefore sensitive to motions in the 10^3–10^6-Hz range.
4. Simulations of the ^{13}C chemical shift anisotropy and ^2H quadrupole echo spectra provide information not only on dynamics, but also on the molecular order of the system. A combination of these two NMR techniques can give a fairly complete picture of the motional, conformational, and structural properties of lipids in model membranes.

Besides phospholipid bilayers consisting only of a single species, the study of mixed systems is particularly fruitful as the different components of the mixed membrane can be observed separately since only one of the different species can be labeled by ^2H or ^{13}C at a time.

This chapter only describes the basic underlying principles and methods of solid state NMR spectroscopy, i.e., those necessary to understand and interpret the line shapes encountered in these studies. A review covering the literature on applications of ^2H and ^{13}C NMR on membranes is not intended. Only selected examples of the different types of spectra observed, and the procedures to extract quantitative information on conformation and dynamics will be given. We will restrict ourselves to lipid bilayers and will not cover the topic of lipid–protein interactions, though this can also be studied by solid state NMR spectroscopy (Oldfield *et al.*, 1978; Kang *et al.*, 1979a,b; Davis *et al.*, 1979a,b; Kang *et al.*, 1981; Rice *et al.*, 1981; Smith and Jarrell, 1983; Davis *et al.*, 1983; Paul, *et al.*, 1985; Lewis *et al.*, 1985).

A detailed treatment of the theory of nuclear magnetic resonance and of solid state NMR techniques with applications to lipid bilayers can be found in textbooks or review articles, respectively (Abragam, 1961; Slichter, 1978; Fukushima and Roeder, 1981; Shaw, 1984; Mehring, 1983; Spiess, 1978; Seelig, 1977; Seelig and Seelig, 1980; Griffin, 1981; Spiess, 1985).

2. PRINCIPLES AND METHODS

Biological membranes and lipid bilayers suspended in water are systems that are not solutions but are lyotropic liquid crystals in which the individual molecules are not able to move isotropically but retain a relatively high degree of molecular order. Thus, when solution NMR techniques are applied, the resulting NMR spectra are broad and featureless, because the orientation-dependent interactions are not averaged by rapid isotropic molecular motion. These orientation-dependent interactions are dipolar interactions, the chemical shift anisotropy, and the quadrupolar interaction. In the following we will describe the basic theory necessary for understanding chemical shift anisotropy spectra and the line shapes encountered in ^{2}H NMR.

2.1. ¹³C Chemical Shift Anisotropy

In liquids the anisotropy of the chemical shift is averaged by the rapid isotropic motions of the molecules in solution and only the "isotropic" part of the chemical shift, which is a scalar quantity, is retained. This isotropic chemical shift arises because the nucleus is shielded from the external magnetic field H_0 by the surrounding electrons. So at the site of the nucleus a different effective magnetic field H_{eff} exists:

$$H_{eff} = H_0(1 - \sigma) = H_0 - \sigma H_0 \qquad (1)$$

The isotropic chemical shift σ expressed in ppm covers a range of ca. 300 ppm for the ^{13}C nucleus, which amounts to ca. 25 kHz at a magnetic field strength of 8.2 tesla.

In solids and semisolids the anisotropy of the chemical shift must also be considered. The shielding of the nuclear moment from the external magnetic field by the surrounding electrons depends now on the orientation of the molecule with respect to the external field. Different orientations will give rise to different resonance frequencies. The anisotropy of the chemical shift is expressed in the form of a second rank tensor, which is diagonal in its principal axis system (PAS). In this coordinate system, which is fixed to the molecular frame in a way not known a priori, the chemical shift tensor σ has the form (Mehring, 1983)

$$\sigma = \begin{pmatrix} \sigma_{11} & 0 & 0 \\ 0 & \sigma_{22} & 0 \\ 0 & 0 & \sigma_{33} \end{pmatrix} \qquad (2)$$

The chemical shift Hamiltonian describing the interactions between the spin system and the external magnetic field has the form

$$\mathcal{H}_{cs} = -\gamma \hbar \mathbf{I} \cdot \sigma \cdot \mathbf{H} \qquad (3)$$

As we are working in the laboratory (LAB) frame, the PAS system of the chemical shift tensor has to be transformed into the LAB system. This can be done by using a series of rotations through the Euler angles ψ, θ, and ϕ, expressed in the form of a rotation matrix (Mehring, 1983):

$$R = R'_x(\psi) R_z(\theta) R_x(\phi) \qquad (4)$$

By applying these rotation matrices to the chemical shift tensor σ_{PAS} in the principal axis system we obtain the shift tensor in the LAB system:

$$\sigma_{LAB} = \mathbf{R}\sigma_{PAS}\mathbf{R}^{-1} \tag{5}$$

Of the resulting tensor σ_{LAB} only the zz element is of interest, because the orientation of the external field is in the z direction of the LAB frame. Only the shielding in this direction is what affects the resonance frequency in the NMR experiment. The zz component of the shift tensor is

$$\sigma_{zz} = \sigma_{11} \sin^2 \theta \cos^2 \phi + \sigma_{22} \sin^2 \theta \sin^2 \phi + \sigma_{33} \cos^2 \theta \tag{6}$$

with σ_{11}, σ_{22}, and σ_{33} being the principal elements of the shift tensor in the principal axis system. For a crystal with only one ^{13}C nucleus we would observe a single line at a resonance frequency depending on the orientation of the crystal in the external magnetic field, the resonance frequency depending on the Euler angles θ and ϕ.

In most cases, however, single crystals are not available and only crystal powders can be investigated where the individual crystallites are randomly oriented. When we assume a uniform distribution of the orientation of the crystallites all solid angles Ω are equally probable and we can determine the intensity I of the NMR signal as a function of the chemical shift (Mehring, 1983):

$$I(\sigma) = |d\Omega/d\sigma| \tag{7}$$

For the general case of an axially asymmetric shift tensor with three different elements the resulting equation for $I(\Omega)$ is relatively complicated. The expression has been calculated by Bloembergen and Rowland (1955) and can also be found elsewhere (Mehring, 1983; Griffin, 1981).

A much simpler case is that of an axially symmetric tensor, i.e., the case where two of the principal elements of the shift tensor are the same, for instance $\sigma_{11} = \sigma_{22} \neq \sigma_{33}$. Equation (6) describing the z component in the LAB frame is then reduced to

$$\sigma_{zz} = (\sigma_{\parallel} - \sigma_{\perp}) \cos^2 \theta + \sigma_{\perp} \tag{8}$$

where we have replaced $\sigma_{11} = \sigma_{22}$ with σ_- and σ_{33} with $\sigma_{\parallel}$ because for $\theta = 0°$, i.e., a parallel orientation of the σ_{33} axis, the "unique" axis of the shift tensor relative to the magnetic field equation (8) leads to

$$\sigma_{zz} = \sigma_{\parallel} = \sigma_{33} \tag{9}$$

and for $\theta = 90°$ we get

$$\sigma_{zz} = \sigma_{\perp} = \sigma_{11} = \sigma_{22} \tag{10}$$

$\Delta\sigma = |(\sigma_{\parallel} - \sigma_{\perp})|$ is referred to as the total anisotropy of the chemical shift tensor. This anisotropy can be quite large. In the general cases of asymmetric

^{13}C shift tensors the differences between σ_{11} and σ_{33} are over 200 ppm for aromatic carbon atoms and are even larger for example for CS_2 (ca. 400 ppm).

The powder line shape for an axially symmetric shift tensor can now be easily calculated from equation (7) with $d\Omega = \sin\theta\,d\theta$. Together with equation (8) one obtains

$$I(\sigma) = \tfrac{1}{2}[(\Delta\sigma)(\sigma - \sigma_-)]^{-1/2} \tag{11}$$

Figures 1a and 1b show the line shapes for the general case and the axially symmetric case, respectively. It can be seen from these figures that the principal elements of the tensor can also be extracted from powder line shapes.

The line shapes discussed above and shown in Fig. 1 apply to crystals, i.e., systems in the rigid limit. Molecular motion will lead to an averaging of the chemical shift. In the extreme of very fast isotropic motions the resulting line shape will be a single sharp Lorentzian line at the position of the "isotropic" chemical shift. If the motions are anisotropic and/or not in the fast limit the resulting line shapes will strongly depend on the type and correlation time of the motions. Thus line-shape analysis can be used to determine the rates and the types of motions of the molecules or of particular segments of larger molecules where we have introduced a ^{13}C nucleus by chemical synthesis.

In the following some examples will be given that illustrate the line shapes observed for different types of motional averaging. We will begin with anisotropic motions in the fast exchange limit, i.e., with motions where the rates are faster than the anisotropy of the chemical shift expressed in Hz $(1/\tau_c \gg \Delta\sigma\gamma H_0)$.

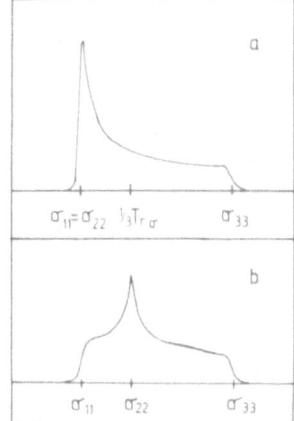

Figure 1. (a) Powder line shape for an axially symmetric chemical shift tensor with $\sigma_{11} = \sigma_{22} = \sigma_\perp$ and $\sigma_{33} = \sigma_\parallel$. (b) Powder line shape for an asymmetric chemical shift tensor with three different components.

2.2. Line Shapes in the Presence of Molecular Motion

2.2.1. Fast Anisotropic Rotation around a Preferred Axis

In the case of fast anisotropic rotation around a preferred axis, called the director axis, this director can have an arbitrary orientation with respect to the principal axis system of the chemical shift tensor. The fast rotation around this axis leads to a new averaged tensor, which by definition is axially symmetric. The averaged tensor components can be calculated by transforming the rigid lattice tensor using rotation matrices into the axis system of the director such that the σ_{33} axis and the director axis are parallel, and then averaging the tensor around the director axis. The new averaged tensor is axially symmetric and has the two principal components $\sigma_\parallel^R$ and $\sigma_\perp^R$, where the superscripts stand for the tensor components averaged by rotation. By these transformations we get (Mehring, 1983)

$$\sigma_\parallel^R = \sigma_{11} \sin^2 \theta \cos^2 \phi + \sigma_{22} \sin^2 \theta \sin^2 \phi + \sigma_{33} \cos^2 \theta \qquad (12)$$

$$\sigma_\perp^R = \tfrac{1}{2}(1 - \cos^2 \phi \sin^2 \theta)\sigma_{11} + \tfrac{1}{2}(1 - \sin^2 \phi \sin^2 \theta)\sigma_{22} + \tfrac{1}{2}\sin^2 \theta \, \sigma_{33}$$

$$= \tfrac{1}{2}[(\sigma_{11} + \sigma_{22} + \sigma_{33}) - \sigma_\parallel^R] = \tfrac{1}{2}[\mathrm{Tr}(\boldsymbol{\sigma}) - \sigma_\parallel^R)] \qquad (13)$$

The averaged anisotropy $\overline{\Delta\sigma^R} = |\sigma_\parallel^R - \sigma_\perp^R|$ will be

$$\overline{\Delta\sigma^R} = \tfrac{1}{2}(3\cos^2 \theta - 1)[\sigma_{33} - \tfrac{1}{2}(\sigma_{11} + \sigma_{22})] + \tfrac{3}{4}(\sigma_{11} - \sigma_{22})\sin^2 \theta \cos^2 \phi \qquad (14)$$

These equations reduce to a much simpler expression if the rigid lattice tensor is axially symmetric to begin with. Then one gets

$$\overline{\Delta\sigma^R} = \tfrac{1}{2}(3\cos^2 \theta - 1)\Delta\sigma \qquad (15)$$

We can now discuss some special cases for different angles θ, the angle between the director axis and the $\sigma_\parallel$ component of the rigid lattice tensor. If $\theta = 0°$, then $\Delta\sigma^R = \Delta\sigma$, the line shape will remain unchanged even though fast axial rotation is present. For a value of 90° $\Delta\sigma^R$ will be $-\tfrac{1}{2}\Delta\sigma$, the line will have half the width of the line shape without motion and will be reversed in sign. For $\theta = 54.7°$, the magic angle, the anisotropy will disappear, we will observe a sharp line at the position of the isotropic chemical shift $\tfrac{1}{3}\mathrm{Tr}(\boldsymbol{\sigma})$, though the motion is still anisotropic. These three different cases are illustrated in Fig. 2. Continuous rotation and jump rotation will lead to the same line shapes in the fast limit, as long as the symmetry is larger than 2.

2.2.2. Fast Discrete Jumps between Two Sites

If the $\sigma_\parallel$ axis of the rigid lattice tensor is executing rapid jumps between different positions the resulting line shape will not be axially symmetric

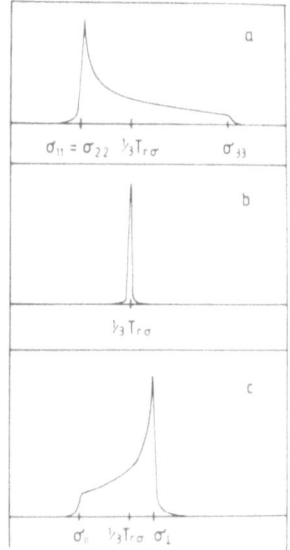

Figure 2. Powder line shapes observed for fast axial diffusion around a preferred axis with different angles θ between σ_{33} and the rotation axis. θ = (a) 0°, (b) 54.7° (magic angle), (c) 90°.

except when the angle between the two positions is 90°. Figure 3 shows the resulting line shapes for different angles β. It should be mentioned that line shapes for angles between 90° and 180° are identical, i.e., angles β and $180° - \beta$ give the same line shape.

2.2.3. Motions in the Intermediate Exchange Regime

In the intermediate exchange regime when $1/\tau_c \approx \gamma \Delta\sigma H_0$ the line shapes are very sensitive to the motional correlation time τ_c, so that the rates of the motions can be extracted from an analysis of the line shape by computer simulations. The calculations of the spectra are based on the well-known problem of chemical exchange. In our case the exchange between different frequencies is caused by the motion of the whole molecule. We will not go into the formalism necessary for the calculation of the line shapes, as this is rather involved. The reader is referred to the literature (Abragam, 1961; Mehring, 1983; Spiess, 1978; Meier, 1983). Figure 4 gives an example for the powder line shapes observed for a twofold jump with an angle of 71° between the two sites. F is the reduced jump rate $2k/\Delta\sigma$, with k being the jump rate constant and $\Delta\sigma$ the anisotropy of the rigid lattice tensor expressed in ω units, i.e., $\sec^{-1}$. The line shapes in Fig. 4 and in Fig. 3 apply only to the case of equal occupation probabilities for the two sites. If these probabilities are not equal, then the line shapes will be different again. Thus the line shapes can be analyzed in terms of the molecular mechanism underlying the motional averaging. Using a plausible

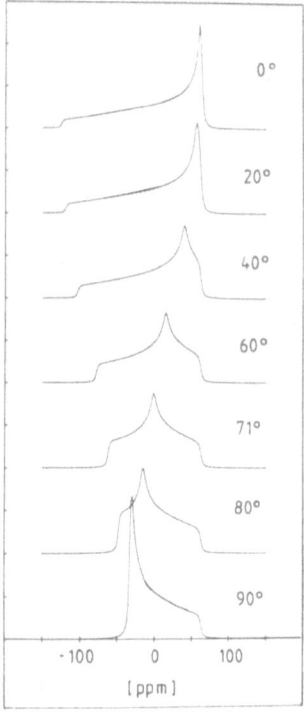

Figure 3. Line shapes for rapid two-site jumps of the σ_{33} axis with different angles β between the two sites.

model for the line-shape calculations one can obtain reliable data for the motional correlation times. It should be kept in mind, however, that different types of motions can result in identical or very similar line shapes. In the fast limit, for instance, a continuous rotation around a director axis with $\theta = 90°$ will lead to the same spectrum as a twofold jump between two discrete sites with an angle of 90° for β (see Figs. 2 and 3). It is therefore important to have independent evidence for the type of motion assumed for the calculation of the spectrum. Additional information on the motional mechanism can come from ^2H NMR spectra, which will be described in the next section.

2.3. ^2H Quadrupole Interaction

^2H NMR spectroscopy is particularly suited for studying molecular order and dynamics in lipid membranes. This was recognized more than ten years ago by Seelig and co-workers, who performed a detailed analysis of the ^2H NMR spectra of specifically labeled lipids in the liquid-crystalline state. These studies gave new insight into the motional behavior and the molecular order of lipids in model and biological membranes (Seelig, 1977;

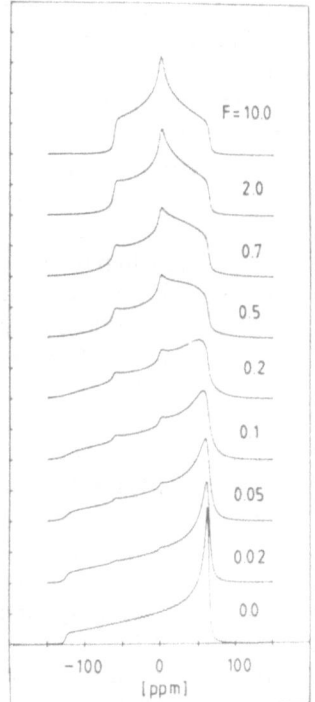

Figure 4. Powder line shapes obtained for a two-site jump motion of the σ_{33} axis of a chemical shift tensor with an angle of 71° between the two sites. F is a reduced jump rate constant: $F = 2k/\Delta\sigma$, with k being the jump rate constant in $\sec^{-1}$ and $\Delta\sigma$ the chemical shift anisotropy.

Seelig and Seelig, 1980; Seelig and Niederberger, 1974; Seelig and Gally, 1976; Seelig *et al.*, 1977).

The application of ²H NMR spectroscopy to lipids in the gel state, however, was not possible at that time because of experimental difficulties in recording the broad powder patterns. In the last years improvements in spectrometer hardware and the application of the quadrupole echo technique made it possible to record undistorted spectra of lipids in the gel phase (Griffin, 1981; Davis *et al.*, 1976; Bloom *et al.*, 1980; Spiess and Sillescu, 1981). Since then ²H NMR spectroscopy has also been applied extensively to study order and dynamics in other liquid-crystalline systems and in polymers (Spiess, 1985; Müller *et al.*, 1985).

Deuterons are ideal nonperturbing nuclear spin labels. They can be incorporated by chemical synthesis into particular sites in the molecule under study. While this is initially a disadvantage, as the synthesis of a specifically labeled molecule can be very complicated, it is also a particular advantage of this method, as different sites in the same molecule can be labeled and then observed separately without interference of resonances arising from other nuclei.

The NMR spectra of deuterons, which have a spin $I = 1$, are dominated by the interaction of the nuclear quadrupole moment with the electric field gradient tensor at the site of the deuteron (Slichter, 1978; Seelig, 1977). This field gradient arises from the electrons surrounding the nucleus. In the case of a C–D bond in an aliphatic chain this electric field gradient is axially symmetric around the bond direction. This interaction is of electrostatic nature; it is strong because of the closeness of the charges, and it is purely intramolecular. Compared to the quadrupole interaction, the chemical shift anisotropy and the dipolar interactions are small.

As deuterons have a spin $I = 1$, $2I + 1 = 3$ different energy levels will be observed, when the nucleus is placed in a magnetic field. In high magnetic fields the quadrupole interaction can still be thought of as a perturbation of the Zeeman interactions. In this case the quadrupole Hamiltonian $\mathcal{H}_Q$ can be written as (Slichter, 1978; Seelig, 1977)

$$\mathcal{H}_Q = \tfrac{1}{4}(eQ/\hbar)\mathbf{I} \cdot \mathbf{V} \cdot \mathbf{I} \tag{16}$$

with e the elementary charge, Q the quadrupole moment of the nucleus (scalar quantity), $\mathbf{I}$ the spin operator, and $\mathbf{V}$ the electric field gradient (EFG) tensor.

The additional perturbation of the Zeeman interaction leads to a change of the three energy levels, so that two transitions with different frequencies ν_+ and ν_-, being symmetric around the unperturbed frequency ν_0, are observed (see Fig. 5). The difference $\nu_+ - \nu_- = \Delta\nu_Q$ is called the quadrupole splitting. $\Delta\nu_Q$ will depend on the orientation of the EFG with respect to the magnetic field. In its principal axis system the EFG tensor can be written as

$$\mathbf{V} = \begin{pmatrix} V_{xx} & 0 & 0 \\ 0 & V_{yy} & 0 \\ 0 & 0 & V_{zz} \end{pmatrix} \tag{17}$$

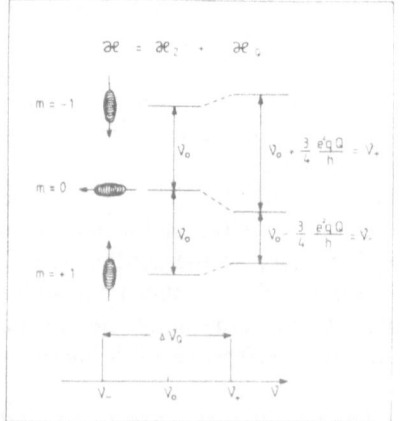

Figure 5. Energy levels for an $I = 1$ spin system like ^{2}H with Zeeman Hamiltonian $\mathcal{H}_z$ and quadrupole Hamiltonian $\mathcal{H}_Q$, giving rise to two different resonance frequencies ν_- and ν_- with a difference $\Delta\nu_Q$, which depends on the angle between the V_{zz} axis of the electric field gradient tensor and the external magnetic field H_0. The situation shown is for a parallel orientation of V_{zz} and H_0.

The EFG tensor is a symmetric and traceless second-rank tensor, i.e.,

$$V_{xx} + V_{yy} + V_{zz} = \text{Tr}(\mathbf{V}) = 0 \tag{18}$$

It is convenient to define two new parameters, namely, the field gradient

$$eq = V_{zz} \tag{19}$$

and the so-called asymmetry parameter:

$$\eta = (V_{xx} - V_{yy})/V_{zz} \tag{20}$$

In the case of an axially symmetric field gradient (aliphatic C–D bond) $\eta = 0$. With the additional definition that $V_{zz} \geq V_{xx} \geq V_{yy}$ it follows that $0 \leq \eta \leq 1$.

As mentioned above, the quadrupole splitting $\Delta\nu_Q$ will depend on the orientation of the EFG with respect to the magnetic field. If the z axis of the EFG is parallel to H_0 we will get a splitting of

$$\Delta\nu_Q(H_0\|z) = \tfrac{3}{2}(eQ/h)V_{zz} = \tfrac{3}{2}(e^2qQ/h) \tag{21}$$

(e^2qQ/h) is referred to as the quadrupole coupling constant. For an aliphatic C–D bond it is roughly 169 kHz, so that the quadrupole splitting for a parallel orientation of H_0 and V_{zz} is 250 kHz. So the separation between the two lines is much larger than the chemical shift anisotropy or the dipolar interaction. In contrast to the chemical shift anisotropy the quadrupole interaction does not depend on the magnetic field strength, so that $\Delta\nu_Q$ is always expressed in Hz and not in ppm.

If V_{zz} is not parallel to the magnetic field we will have to perform a coordinate transformation of the EFG tensor from its PAS system into the LAB frame. As described before for the chemical shift anisotropy [see equations (4)–(6)], we will get

$$\Delta\nu_Q(\theta, \phi) = \tfrac{3}{2}(e^2qQ/h)[(3\cos^2\theta - 1)/2 + \tfrac{1}{2}\sin^2\theta \cos 2\phi] \tag{22}$$

For an aliphatic C–D bond with $\eta = 0$ this reduces to

$$\Delta\nu_Q(\theta, \phi) = \tfrac{3}{2}(e^2qQ/h)\tfrac{1}{2}(3\cos^2\theta - 1) \tag{23}$$

We thus have a similar angular dependence of the quadrupole splitting as described before for the chemical shift anisotropy [see equation (8)].

In a crystal powder all orientations of the unique axis of the EFG tensor with respect to the magnetic field are equally probable. We have to sum over all solid angles to get the powder pattern. As we have two individual transitions the final powder pattern will be the sum of two overlapping powder patterns of similar shape but reversed sign as shown in Fig. 6. The inner two intense peaks correspond to a perpendicular orientation of V_{zz}

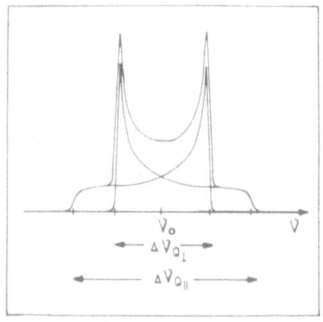

Figure 6. Powder line shape for an $I = 1$ spin system, being a superposition of two powder patterns of reversed sign for the $m = +1 \to 0$ and the $m = 0 \to -1$ transition. The outer edges correspond to $\Delta \nu_{Q_\perp}$ while the inner peaks are separated by $\Delta \nu_{Q}$.

with respect to the magnetic field. The separation of these two peaks is

$$\Delta \nu_{Q_\perp} = \tfrac{3}{4}(e^2 qQ/h) = 125 \text{ kHz} \tag{24}$$

and the separation of the outer edges of the powder pattern corresponds to a parallel orientation:

$$\Delta \nu_{Q_\parallel} = \tfrac{3}{2}(e^2 qQ/h) = 250 \text{ kHz} \tag{25}$$

Because of the enormous width of the rigid limit powder pattern the free induction decay (FID) after a normal 90° pulse will decay very rapidly with a time constant corresponding roughly to the inverse of the total width of the spectrum, i.e., a few microseconds. This poses a severe problem, as the recording of the FID has to start immediately after the 90° pulse. However, this is not possible for technical reasons, namely, receiver overload in the first few microseconds after the pulse. Another complication arises from the necessity to get a uniform power distribution over the whole spectral width so that nuclei resonating at different frequencies will all be flipped by the same angle by the radiofrequency pulse. This requires high transmitter power to get short 90° pulses. If the pulses are longer, severe intensity decreases will result at frequencies at the outer edges of the spectrum (Griffin, 1981; Bloom *et al.*, 1980). To overcome the problems of having to record the extremely short FID where receiver overload in the first microseconds after the pulse will prevent the recording of the true intensity, a quadrupole echo pulse sequence is used (Davis *et al.*, 1976; Solomon, 1958). This sequence consists of a pair of 90° radiofrequency pulses whose phases differ by 90° (see Fig. 7). Between the two pulses a waiting time τ of 20–100 μsec is inserted. The maximum of the echo is observed at a time 2τ. This echo can now be recorded without distortion, as the receiver has recovered from the rf pulses. The echo is symmetric around the maximum. Beginning at the peak of the echo it is Fourier transformed to give the spectrum.

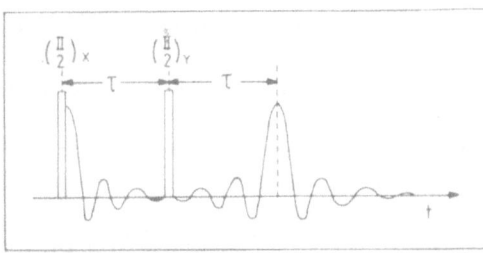

Figure 7. Quadrupole echo pulse sequence consisting of two $\pi/2$ pulses with different phases x and y. The separation τ between the two pulses is normally varied between 20 and 100 μsec. The Fourier transformation giving the spectrum is started at time 2τ at the top of the echo.

2.4. Line Shapes in the Presence of Molecular Motion

2.4.1. Fast Anisotropic Rotation around a Preferred Axis

In the presence of molecular motions the line shape will change. Particularly simple cases are those where the motional correlation times are very short, i.e., the fast limit case. The same equations discussed above for the chemical shift anisotropy also apply to the deuterium spectra, only the chemical shift anisotropy σ has to be replaced by the quadrupole interaction. For fast anisotropic rotation around the director axis the splitting will be reduced according to

$$\overline{\Delta \nu_{Q^R}} = \Delta \nu_Q \tfrac{1}{2}(3 \cos^2 \theta - 1) \tag{26}$$

with θ being the angle between the C–D bond and the rotation axis. Thus we will get the same line shapes as shown in Fig. 2 for the single transitions, only for deuterium we will have to superimpose two lines of opposite sign. Specifically for $\theta = 90°$ we will get a powder pattern of identical shape as in Fig. 6 but reduced in width by a factor of 2, and for the magic angle we will observe a single sharp line at ν_o.

2.4.2. Fast Discrete Jumps between Different Sites

For discrete jumps we will get the same line shapes as shown in Fig. 3 for the individual transitions. Again two of them with opposite sign have to be superimposed to get the deuterium spectra.

2.4.3. Line Shapes in the Intermediate Exchange Regime

When the reorientation rates become smaller we are approaching the intermediate exchange regime. The line shapes are then very sensitive to the rates of the reorientational motions. This is the same situation as described before for the chemical shift anisotropy spectra, where line shapes for a two-site jump motion were shown in Fig. 4. For the chemical shift anisotropy of ^{13}C the intermediate exchange regime is approached when

$1/\tau_c \approx \gamma H_0 \Delta\sigma$, so it depends on the magnetic field strength and on the anisotropy. For a field strength of ca. 7 T and an anisotropy of 150 ppm this occurs when $\tau_c \approx 1.4 \times 10^{-5}$ sec. For deuterium the intermediate exchange regime is approached at shorter correlation times, as the width of the rigid limit powder pattern is larger by about one order of magnitude. Thus ^{2}H spectra are particularly sensitive to motions with correlation times between 10^{-7} and 10^{-5} sec.

An additional effect arises from the application of the quadrupole echo sequence (see Fig. 7). When the motional correlation times are comparable to the pulse spacing τ in the quadrupole echo sequence the line shapes will depend on the exact value of τ. In addition, strong decreases in echo intensity, i.e., integral intensity of the powder pattern, will occur, as some spectral components will have irreversibly dephased and cannot be refocused by the second 90° pulse of the echo sequence (Spiess and Sillescu, 1981). These effects can be exploited to differentiate between different motional mechanisms, which in the fast limit give similar line shapes, because in the intermediate exchange regime the τ dependence of the line shape will differ. An example of this effect is shown in Figs. 8 and 9 for two types of motions, namely, a two-site 90° jump of a C–D vector and a threefold rotational jump with the C–D vector oriented perpendicular to the rotational jump axis. In the fast limit these two motions give identical line shapes. Even for a jump rate constant k of 3×10^6 sec^{-1} the powder patterns are almost indistinguishable when the pulse spacing is set to 10 μsec. Increased pulse spacing τ, however, leads to different line shapes, so that a distinction between these two different motional mechanisms is possible. Note also that the integral intensity decreases with increasing τ, but that this decrease is also different for the two types of motion.

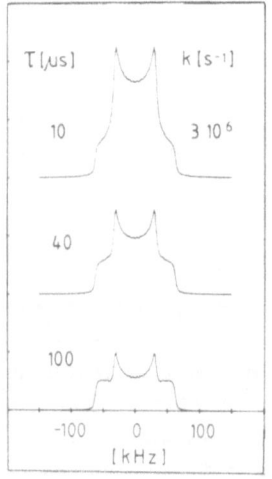

Figure 8. Dependence of the powder line shape for a two-site 90° jump of a C–D vector on the pulse spacing τ in the quadrupole echo sequence (k is a jump rate constant).

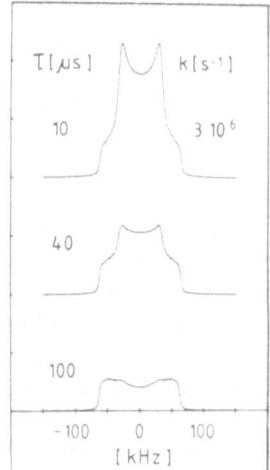

Figure 9. Dependence of the powder line shape for a threefold rotational jump of a C–D vector oriented perpendicular to the rotation axis on the pulse spacing τ of the quadrupole echo sequence.

3. APPLICATIONS OF ¹³C AND ²H NMR TO LIPID MODEL MEMBRANES

3.1. ¹³C NMR of Phospholipids

¹³C natural-abundance NMR spectroscopy has been used quite extensively to characterize the behavior of lipids in the liquid-crystalline phase above the phase transition temperature. For lipids in the liquid-crystalline phase high-resolution spectra can be obtained when small sonicated vesicles are used (Levine *et al.*, 1972a,b). This is so because fast lateral diffusion of the lipids around the vesicle surface in addition to fast tumbling of the whole vesicle leads to almost isotropic averaging of the chemical shift anisotropy. When the vesicles are cooled below T_m the resolution decreases and the spectrum is almost completely lost. The reason for this is that lateral diffusion has slowed down by at least two orders of magnitude. In addition vesicle aggregation and fusion prevents fast vesicle tumbling, so that motional averaging is no longer complete.

For lipids in the liquid-crystalline phase T_1 relaxation times of the various carbon atoms of the methylene segments and of the glycerol backbone and the head group have been determined by Levine *et al.* (1972a,b). These data show that the mobility of the chain segments increases with larger distance from the glycerol backbone, the last two to three segments displaying considerably longer relaxation times. Thus the mobility is highest at the end of the aliphatic chains and lower in the middle part of the chain. Relaxation times can be calculated using the assumption that the molecules are rotating very fast about their long axis and that *trans-gauche* isomeriz-ation of the chains is also fast with correlation times of 10^{-10} to 10^{-9} sec (Pace and Chan, 1982a). In the liquid crystalline phase the lipids are thus highly mobile, displaying not only fast rotational diffusion and *trans-gauche* isomerization of the chains, but also lateral diffusion with diffusion coefficients of 10^{-8}-10^{-7} cm^2/sec (Sackmann and Traüble, 1972; Devaux and McConnell, 1972; Pace and Chan, 1982b).

All these motions are considerably slowed down when the phos-pholipids are cooled below the phase transition temperature into the gel state. As mentioned above, this leads to the loss of the high-resolution spectrum. With sufficient ^1H decoupling power to suppress the dipolar interactions between the protons and the ^{13}C carbon atoms powder line shapes of all carbon atoms in the lipid molecules can be obtained. However, because the chemical shift anisotropy of the methylene carbon atoms is larger than the differences in the isotropic chemical shift, all resonances from the methylene carbons overlap. Thus a line-shape analysis of powder patterns of individual methylene carbons in the chains, the backbone, or the head group is not possible. High resolution can be regained by applying the magic angle spinning technique (MAS) (Griffin, 1981). However, the information on the motional behavior contained in the powder line shapes is then lost.

The only resonances in the ^{13}C solid state spectra that do not overlap are the lines of the C=O carbons of the two ester groups of the glycerol backbone. This offers the possibility of studying the powder patterns of this particular site in the molecule and applying line-shape analysis. However, natural abundance ^{13}C NMR is not particularly well suited for this purpose, because of the long accumulation times necessary to get a good signal-to-noise ratio and the fact that there are two ester carbon atoms in a molecule whose resonances overlap and which can behave differently (Cornell, 1981). A way to circumvent this problem is to chemically label one of the carbonyl ester atoms with pure ^{13}C so that the intense resonance observed in the carbonyl region of the spectrum can now be unequivocally assigned to the labeled C=O group (Wittebort *et al.*, 1981).

^{13}C NMR experiments with phospholipids labeled at the C=O group of the *sn*-2 chain have been carried out with quite a number of systems (Wittebort *et al.*, 1981; 1982). Figure 10 shows spectra of 2(1-^{13}C)DPPE at different temperatures below and above the phase transition temperature. The bottom trace is a spectrum of the lipid at −55°C, which is identical to a spectrum of the dry lipid powder (Rice *et al.*, 1981; Wittebort, 1982; Blume *et al.*, 1982a). The line on the left-hand side of the spectrum is the one of the labeled C=O group. All other lines arise from natural abundance ^{13}C carbon atoms in the chains, the head group, and the backbone. The line shape of the ^{13}C=O carbon indicates that the chemical shift tensor is not axially symmetric, but has three different values $\sigma_{11} = -133$, $\sigma_{22} = -13$,

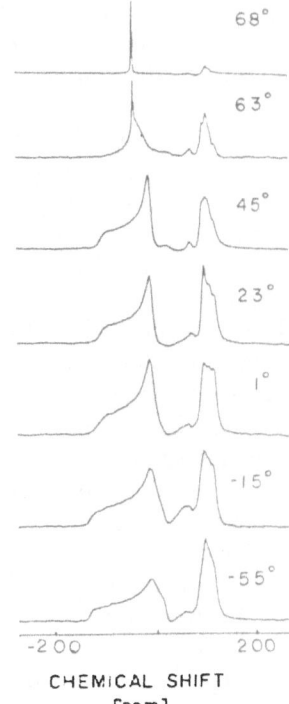

Figure 10. ¹³C spectra of 2(1-¹³C)DPPE at different temperatures. The powder line shape on the left-hand side arises from the labeled C = O group of the *sn*-2 chain. Resonances on the right are due to natural abundance of methylene carbon atoms in the chains and the head group (Blume *et al.*, 1982a).

and $\sigma_{33} = +11$ ppm relative to external benzene. However, when the bilayers are brought to room temperature and above, the line shape transforms into an axially symmetric one of reduced width. At 45°C we find only two principal values $\sigma_\parallel = -110$ and $\sigma_\perp = -15$ ppm. Thus the anisotropy σ is reduced to 95 ppm compared to the rigid limit value of 146 ppm. Above 64°C, the phase transition temperature of DPPE, the spectrum collapses to a narrow line of ca. 2 ppm width.

These temperature-dependent changes of the line shape can be explained using the following assumptions. At −55°C and in the dry powder the motion of the molecules is frozen out on the ¹³C time scale so that the rigid lattice powder pattern is observed. With increasing temperature the molecules begin to execute rotational jumps in the quasihexagonal lattice so that the asymmetric chemical shift tensor begins to get averaged. At high temperature just below the phase transition temperature this motion is fast compared to the anisotropy. This rotational jump motion produces an axially symmetric powder pattern of reduced width. As the rigid lattice tensor is asymmetric and the orientation of its principal axis system with respect to the molecular frame is not known, the orientation of the ester group relative to the rotational jump axis cannot be determined. For simplification we can assume an axially symmetric rigid lattice tensor with an anisotropy of 146 ppm. Then we can determine the orientation of the unique tensor axis relative to the rotation axis according to equation (15). Using a value of 97 ppm for the averaged tensor, we find that the angle θ is ca. 27° at a temperature of 25°C (Wittebort et al., 1982). The narrow line observed in the liquid-crystalline phase can be explained by the assumption of a slight conformational change at the ester carbonyl so that the unique tensor axis is now close to the magic angle (54.7°). This is the simplest assumption that would produce the observed effect. However, as we know from other NMR experiments, the mobility and internal flexibility of the molecules is high. Thus it is more likely that some type of internal motion in addition to fast long-axis rotation is present, which leads to an averaging process around the magic angle.

The spectra taken below T_m in the gel phase indicate a fast motional averaging process with a correlation time shorter than 10^{-4}–10^{-5} sec producing an axially symmetric powder pattern. The assumption of a rotational jump motion in the quasihexagonal gel phase was the simplest assumption to describe the line shape. It should be mentioned that other motions, e.g., a 90° two-site jump, can also produce this line shape (see Fig. 3). Because of the hexagonal symmetry of the DPPE gel phase this motion seems unlikely.

The ^{13}C spectra of other phospholipids labeled at the same position, namely, the carbonyl group of the sn-2 chain show similar line shapes. Figure 11 shows spectra of 2(1-^{13}C)DPPC at different temperatures (Wittebort et al., 1981; Blume et al., 1982b). The phase behavior of phosphatidylcholines is more complicated than that of phosphatidylethanolamines as different types of gel phases are observed. In particular phosphatidylcholines with two chains of identical length display a characteristic rippled $P_{\beta'}$ or P_β phase just below the main transition to the liquid crystalline L_α phase (Janiak, 1976). In this ripple phase the surface of the bilayers is distorted by periodic undulations ca. 150–300 Å apart. The exact structure of this phase is still not completely clear, as the orientation of the chains with respect to the average bilayer surface is not known. In addition there is still some dispute whether the surface is distorted by a sinusoidal or a more sawtooth-like pattern (Rüppel and Sackmann, 1983; Stamatoff et al., 1982; Gebhardt et al., 1977; Larsson, 1977; Doniach, 1979; Meier et al., 1983). The ^{13}C NMR data as well as the deuterium spectra (see below) indicate that this ripple phase is microscopically heterogeneous as two distinctly different components can be observed in the temperature range covering the pretransition of DPPC at 35°C (see Fig. 11). In addition the relative intensities of the two components change with temperature and the line shapes also indicate an exchange between these two components. It has been proposed that the two components could indicate different

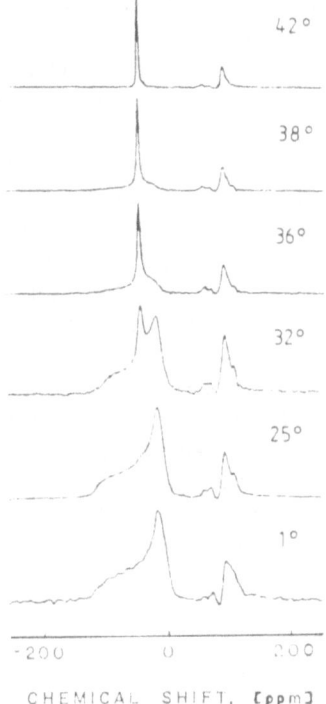

Figure 11. ¹³C spectra of 2(1-¹³C)DPPC at temperatures covering the pre- and main transition region. A superposition of a narrow line at the isotropic chemical shift with a wide axially symmetric powder pattern is observed in this temperature range (Blume et al., 1982b).

conformations of the DPPC molecules on the convex and concave side of the ripples and that lateral diffusion perpendicular to the ripples leads to this conformational change of the DPPC (Wittebort et al., 1982). For DPPC the pretransition occurs at 35°C and the main transition at 41°C. It can be seen from Fig. 11 that the sharp line characteristic of liquid-crystalline lipid appears at temperatures well below the pretransition temperature (see the 32°C spectrum). Thus the conformational change in the DPPC molecules affecting the ¹³C spectrum takes place over a temperature range of ca. 15°C below the main transition and is essentially complete at 41°C.

In the low-temperature $L_{\beta'}$ phase of phosphatidylcholines the ¹³C spectra look very similar to the spectra of phosphatidylethanolamines (see Fig. 10). Rotational diffusion persists down to temperatures somewhat below the freezing point of water. However, when DPPC is incubated at 0°C for several days the $L_{\beta'}$ phase transforms into a highly ordered subphase with a rectangular unit cell called the L_σ phase (Chen et al., 1980; Ruocco and Shipley, 1982; Church et al., 1986). The transition between these two gel phases is commonly called the subtransition. In agreement with the high amount of order found by X-ray diffraction, the ¹³C spectrum of DPPC in the L_σ phase displays an asymmetric line shape for the ¹³C=O group identical to the spectrum observed for the dry powder. Thus in the L_σ phase there is no motion on the ^{13}C time scale (Lewis *et al.*, 1984).

Similar results have been obtained for mixed-chain phosphatidylcholines, which convert directly from the rippled phase to the L_σ phase upon decreasing the temperature (Lewis *et al.*, 1984). Though the ^{13}C spectra indicate no motion in the glycerol backbone part of the molecule, that does not mean that the molecules are completely rigid. Deuterium spectra of chain labeled compounds show that even at low temperatures there is still some *trans-gauche* isomerization in the chains (Lewis *et al.*, 1984).

Glycolipids like galactosylcerebrosides have large well-hydrated head groups and many hydroxyl groups at the bilayer water interface capable of intermolecular hydrogen bonding. The ^{13}C spectra of these glycolipids labeled at the 1-position of the fatty acyl chain, which in cerebrosides is linked by an amide bond to the sphingosine residue, are distinctly different from those of phospholipids. Figure 12 shows spectra of 1-^{13}C-palmitoylgalactosylcerebroside at different temperatures (Griffin, 1981). The line shape observed for the dry powder shows that the chemical shift tensor is asymmetric. This line shape is retained for the hydrated lipid up to temperatures close to the phase transition temperature at ca. 80°C. In the

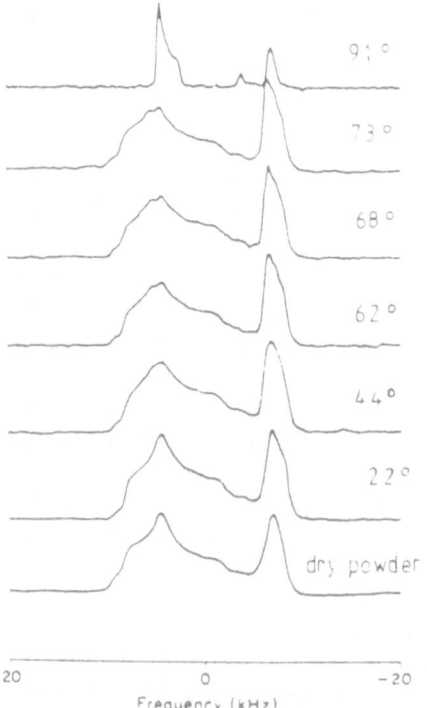

dry powder

20   0   −20

Frequency (kHz)

Figure 12. ^{13}C spectra of N-palmitoyl-galactosylcerebroside labeled at the amide carbon atom. The rigid lattice powder pattern observed for the dry powder is preserved in the hydrated system almost all the way up to the transition temperature of ca. 80°C (Griffin, 1981).

liquid crystalline phase the spectrum is narrowed and the line shape becomes axially symmetric, indicating fast rotational diffusion of the molecules. Thus glycolipids show essentially only very slow rotational diffusion in the gel state, presumably because of the intermolecular hydrogen bond network between head groups and between the sphingosine backbones.

^{13}C solid-state NMR can also be used to study the behavior of lipid mixtures and to construct phase diagrams for these systems (Blume *et al.*, 1982b; Blume and Ackerman, 1974). This method utilizes the fact that for phospholipids in the liquid crystalline phase a very narrow line is observed (see above). For instance, the 63°C spectrum of DPPE in Fig. 10 is composed of a superposition of a sharp line and an axially symmetric powder pattern, though the line shape indicates exchange between these two components. This superposition originates from the existence of gel and liquid-crystalline domains that are in equilibrium over the temperature range of the DPPE transition, which has a half-width of ca. 1.5°C (Blume and Ackerman, 1974; Blume, 1980). For binary mixtures in which the transition temperatures of the two components differ by, for instance, 20°C, the transition region is much broader. Consequently the temperature interval in which coexistence of both phases is observed is larger. An example of this case is mixtures of DPPC with DPPE. In this binary system the components differ only in the structure of the head group, while the chain lengths are the same. Because of its stronger head group interactions DPPE has the higher transition temperature (64°C) compared to DPPC (41°C). Mixtures of these two lipids show complete miscibility in both the gel and the liquid-crystalline phases, though the mixing behavior is nonideal. Figure 13 shows spectra of a 1:1 mixture of DPPE with DPPC at different temperatures covering the phase transition region. On the left-hand side of this figure DPPC is the labeled species, while spectra on the right show the same mixture, only with DPPE being the labeled compound. The phase transition region is between 45°C and 55°C. Clearly the coexistence of liquid-crystalline and gel state domains can be seen as superpositions of axially symmetric powder patterns with sharp lines characteristic of the L_α-phase. As DPPC is the lipid with the lower transition temperature it should be enriched in the liquid-crystalline phase. This effect is quite evident when the spectra on the left- and right-hand sides of Fig. 13 are compared. At 50°C, for instance, the intensity of the sharp line in the 2(1-^{13}C)DPPC spectrum on the left is higher than that in the 2(1-^{13}C)DPPE spectrum on the right.

The fact that these lines are only a little affected by exchange indicates that the "fluid" and "solid" domains must be large, so that the exchange rates are slow on the ^{13}C time scale. Indeed, in electron microscopy and electron diffraction large domains of at least 200–500 nm diameter can be distinguished (Sackmann, 1978; Hui, 1981). Because these domains are so large, only molecules in the vicinity of the domain boundary can exchange

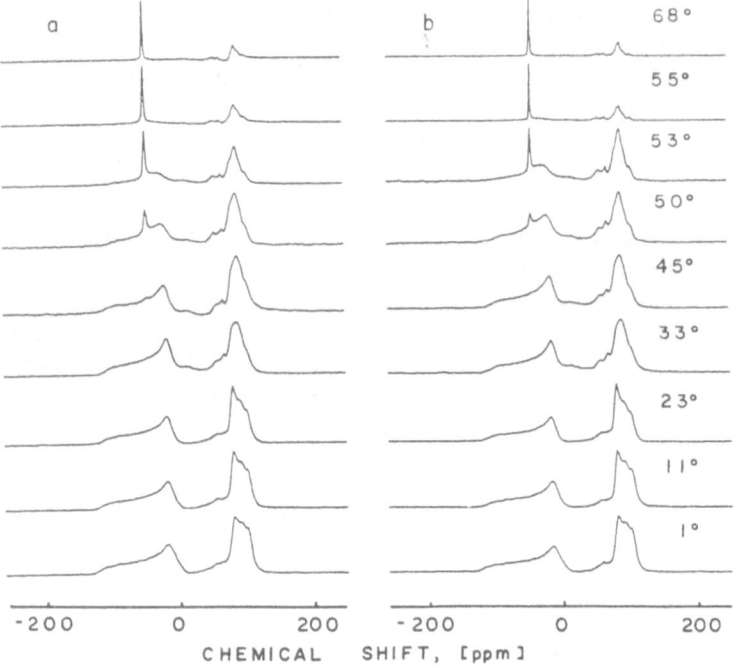

Figure 13. ^{13}C spectra of a 1:1 DPPC–DPPE mixture: (a) $2(1\text{-}^{13}C)DPPC\text{-}DPPE$, (b) DPPC–$2(1\text{-}^{13}C)DPPE$. Note that spectra in the phase transition region (ca. 45–55°C) consist of a superposition of liquid-crystalline-like lines with axially symmetric powder patterns (Blume *et al.*, 1982b).

on the ^{13}C time scale. The domain sizes as well as the exchange rates, which are related to the diffusion coefficients in the gel and liquid-crystalline phase, are temperature dependent. So spectra at higher temperatures indicate more rapid exchange. Likewise the composition of the mixture influences the domain sizes. Spectra of 1:3 DPPC/DPPE mixture show an increased exchange broadening (not shown) (Blume *et al.*, 1982b).

The ^{13}C spectra of these mixtures can also be analyzed for structural and conformational changes occurring in the gel phase. At 25°C the width of the $2\text{-}^{13}C{=}O$ powder pattern is different for DPPC and DPPE. Figure 14 shows the chemical shift anisotropy for DPPC/DPPE mixtures as a function of composition. A continuous decrease is observed with increasing DPPE content, the anisotropies being the same for both components of the mixture. The latter result indicates that we are dealing with a homogeneous mixture with no domain formation. Otherwise we would observe the anisotropies of the pure components. In addition the continuous change of the width of the powder pattern can be interpreted as a change of the tilt

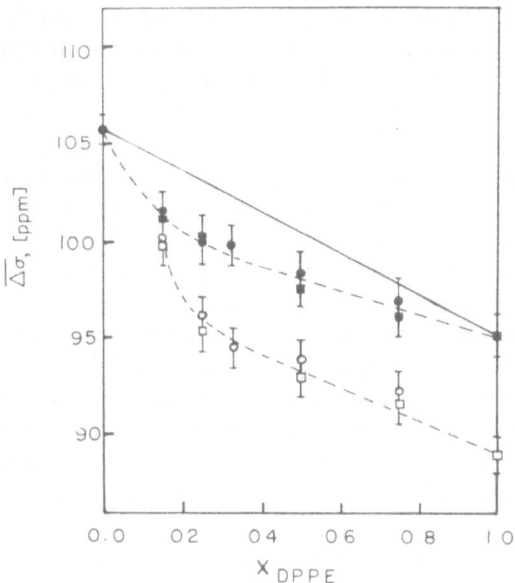

Figure 14. Residual chemical shift anisotropy $\overline{\Delta\sigma}$ of the sn-2 ^{13}C = O powder patterns in DPPC-DPPE mixtures as a function of composition. ●, 2(1-^{13}C)DPPC at 25°C; ○, at $T = T_{\text{onset}}$; ■, 2(1-^{13}C)DPPE at 25°C; □, at $T = T_{\text{onset}}$ (Blume *et al.*, 1982b).

angle of the acyl chains. DPPC at 25°C has a chain tilt of ca. 30° with respect to the bilayer normal owing to its large head group (Janiak *et al.*, 1976; Tardien *et al.*, 1973). In contrast, DPPE has no chain tilt (McIntosh, 1980). For the mixtures this tilt angle seems to decrease with increasing DPPE concentration, but not in a linear fashion. At low DPPE content the decrease is stronger. Apparently a gradual transformation from an $L_{\beta'}$ to an L_{β} phase is taking place, this change being essentially complete at ca. 50 mol% DPPE (Blume *et al.*, 1982b).

Another lipid mixture that has been analyzed in detail is the DPPE-cholesterol system (Blume and Griffin, 1982). Cholesterol is present in most eukaryotic membranes. The effects of cholesterol incorporation into model membranes and its biological function have been studied in great detail (Ladbrooke *et al.*, 1968; Demel and Kruyff, 1976; Estep *et al.*, 1978; Mabrey *et al.*, 1978). However, there is still some uncertainty whether cholesterol induces phase separation in the gel state and whether phase separation into fluid and solid state domains occurs (Rubinstein *et al.*, 1979; Lentz *et al.*, 1980) at temperatures inside the range of the endothermic transition observed by differential scanning calorimetry (DSC) at intermediate cholesterol levels (20–40 mol%). The effect of cholesterol on the thermotropic transition of phospholipids as seen by DSC is first a broadening of the phase transition at low cholesterol content, and finally at 50 mol% a complete disappearance of the endothermic transition (Ladbrooke *et al.*, 1968; Estep *et al.*, 1978; Mabrey *et al.*, 1978). At intermediate cholesterol concentrations the calorimetric peaks are similar in width to those for binary mixtures of phospholipids but seem to be a superposition of a sharp endotherm and a broad less cooperative peak. In binary phospholipid mixtures phase separation can be observed in the temperature region of the transition (see above).

The ^{13}C spectra of $2(1-^{13}C)$DPPE/cholesterol mixtures are shown in Fig. 15. They indicate a thermotropic behavior distinctly different from that of phospholipid mixtures. As the cholesterol concentration is increased from 10 to 50 mol% the narrow L_α-like line appears at progressively lower temperature, and there always exists a temperature range where two-component spectra are observed. In contrast to the spectra for DPPC–DPPE mixtures these two-component line shapes are broadened owing to an exchange process with exchange rates comparable to the chemical shift anisotropy.

The columns on the right-hand side of Fig. 15 show computer simulations assuming an exchange process between one component with a sharp line at the isotropical chemical shift being characteristic for liquid-crystalline lipid, and another component giving the axially symmetric powder pattern observed for phospholipids in the gel state (Blume and Griffin, 1982). The parameters that are varied are the exchange rate and the fraction of liquid-crystalline-like lipid, f_L. As can be seen from Fig. 15, this simple model reproduces the experimental spectra quite well. In Fig. 16 the fraction f_L is plotted versus temperature and compared to the normalized transition curves obtained from DSC experiments on the same system (Blume, 1980). It is obvious that the fluidlike fraction obtained from the spectra does not correspond to the degree of transition obtained by calorimetry. From the shape, i.e., the broadening of the two-component spectra, it is also evident that in the case of DPPE/cholesterol no large domains of fluid and solid lipids coexist. The ^{13}C spectra can be interpreted as follows: Cholesterol is incorporated into DPPE bilayers in an almost homogeneous fashion, at least up to concentrations of ca. 30 mol% and temperatures not below 0°C (Blume and Griffin, 1982). The L_α-like line observed at low temperatures below the DSC endotherms could arise from those molecules being perturbed, because they are in contact with cholesterol molecules. These molecules exchange via lateral diffusion, with unperturbed lipid molecules being farther away from the cholesterol. When the temperature is increased the lipid bilayers "melt" more or less cooperatively and all molecules are then transformed to a more liquid-crystalline state. We will see from the results of deuterium NMR experiments that this liquid-crystalline state in mixtures with cholesterol is different from the L_α-phase of pure phospholipids (see

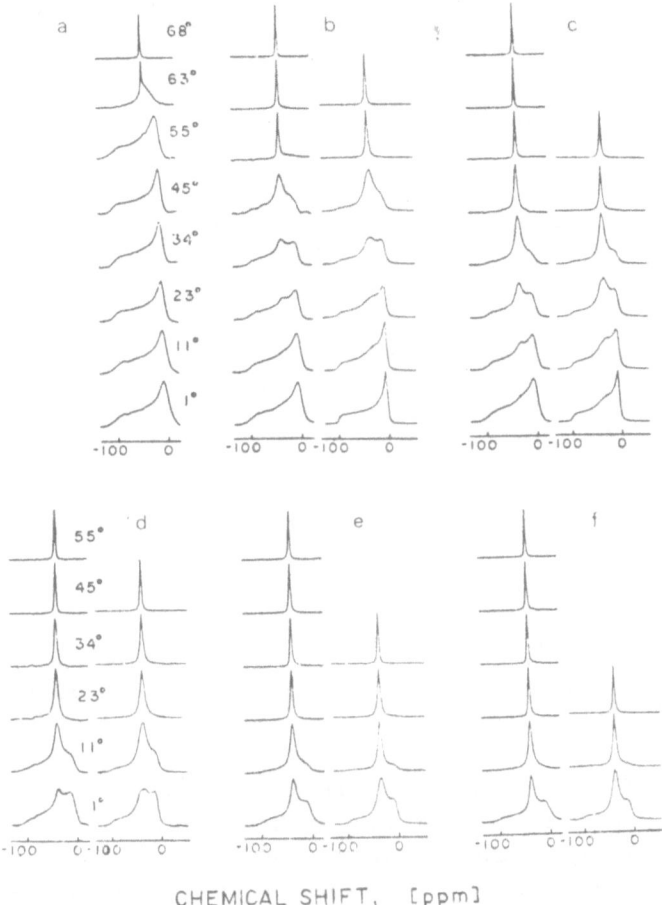

CHEMICAL SHIFT, [ppm]

Figure 15. ¹³C spectra of 2(1-¹³C)DPPE–cholesterol mixtures. (a) pure DPPE; (b) 10 mol%; (c) 20 mol%; (d) 30 mol%; (e) 40 mol%; (f) 50 mol% cholesterol. On the right-hand side next to the experimental spectra are computer simulations of the line shapes using the model described in the text. Only the ¹³C=O resonances are shown (Blume and Griffin, 1982).

below). An important result of the ¹³C NMR experiments is that in phospholipid/cholesterol mixtures the exchange between the two different components is fast. This means that there is no phase separation into larger domains of different composition and physical state. If there are domains at all, they must be very small, consisting only of ca. 100 molecules or less. These results are in agreement with X-ray and electron diffraction results, which show no domain formation in phosphatidylcholine–cholesterol mixtures (Hui and He, 1983). Recent kinetic experiments also support this view (Blume and Hillmann, 1986).

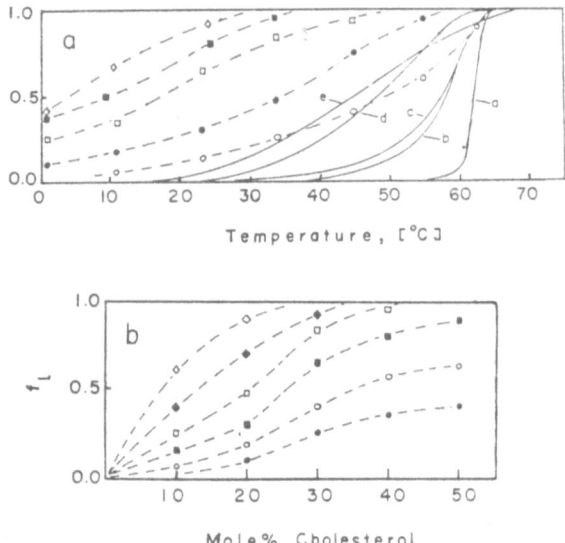

Temperature, [°C]

Mole% Cholesterol

Figure 16. (a) Fraction of liquid-crystalline-like lipid f_L obtained from the spectral simulations shown in Fig. 15 versus temperature: ○, 10; ●, 20; □, 30; ■, 40; ◇, 50 mol% cholesterol, respectively. The solid lines are the normalized transition curves obtained from DSC measurements (a) 0, (b) 10, (c) 20, (d) 30, (e) 40 mol% cholesterol (Blume, 1980; Blume and Griffin, 1982). (b) f_L plotted versus cholesterol content at different temperatures: ●, 1°C; ○, 11°C; ■, 23°C; □, 34°C; ◆, 45°C; ◇, 55°C.

^{13}C NMR of lipids specifically labeled at the carbonyl carbon atom of one of the acyl chains is an ideal method to determine whether lipid molecules are in the liquid-crystalline or the gel state. In addition, the line shape in the gel state indicates whether or not fast axial diffusion exists. So dynamical and structural information on pure lipids and lipid mixtures can be obtained. Because the chemical shift anisotropy for this carbon atom is ca. 10–15 kHz at a field strength of ca. 7 T, the spectra are sensitive to motions with correlation times down to 500 sec^{-1}. The combination of ^{13}C chemical shift anisotropy spectra with deuterium quadrupole spectra thus covers a dynamical range of 4–5 orders of magnitude, a time range where most exchange processes in gel state lipid bilayers occur.

3.2. ^{2}H NMR of Phospholipids

^{2}H NMR spectroscopy requires the labeling of the lipid with a deuterium nucleus at a particular site. Much simpler is the perdeuteration of the acyl chains. However, the spectra are then composed of overlapping powder patterns of all the CD_2 groups in the acyl chains, and this complicates the quantitative analysis. Specific labeling of certain positions in the chains or the head group circumvents this problem. Though the chemical synthesis is considerably more complicated. Deuterium NMR has been used quite extensively to characterize the behavior of model membranes and of biological membranes. As we will only describe a number of examples, the reader is referred to several reviews that have appeared on this subject (Smith and Jarrell, 1983; Seelig, 1972; Seelig and Seelig, 1980; Griffin, 1981; Smith, 1979; Jacobs and Oldfield, 1981; Smith and Oldfield, 1984).

3.3. Phospholipids in the Liquid-Crystalline Phase

²H spectra of phospholipids in the liquid-crystalline state are relatively easy to obtain because the quadrupole splittings are small compared to the rigid lattice powder patterns. Almost all motions executed by the molecules are in the fast limit of the ²H NMR time scale, so that only limited information can be extracted by a line-shape analysis. The fast limit powder patterns are therefore commonly interpreted in terms of the order parameter concept (Seelig, 1977). If a Cartesian coordinate system x, y, z is fixed to a particular group in a molecule, then the order parameter S_{ii} can be written as

$$S_{ii} = \tfrac{1}{2} \overline{(3 \cos^2 \theta_i - 1)}; \qquad i = x, y, z \qquad (27)$$

The order parameters describe the average fluctuations of the x, y, and z axes with respect to a preferred axis, called the director D, and $\overline{\cos^2 \theta_i}$ is the time average of the angular fluctuation of the ith coordinate axis with θ_i being the angle between the particular coordinate axis and the director D. From the orthogonality relations of the direction cosines it follows that $\sum_i S_{ii} = 0$, so that we have only two independent order parameters. In the case of ²H spectra of a CD_2 group the situation is even simpler as the field gradient tensor of a C–D bond is nearly axially symmetric, i.e., $\eta = 0$. Therefore only one order parameter remains which can be determined and which is commonly called S_{CD}. The quadrupole splitting $\overline{\Delta \nu_Q}$ observed is then related to the order parameter S_{CD} by

$$\overline{\Delta \nu_{Q_-}} = \tfrac{3}{4}(e^2 qQ/h)S_{CD} \qquad (28)$$

Usually the segmental order parameter S_{mol} of a CD_2 group in a methylene chain is used as a measure of order, where the segment direction is defined by the normal to the plane spanned by the CD_2 group (Seelig, 1977). These two order parameters are then related by

$$S_{mol} = -2S_{CD} \qquad (29)$$

Seelig and Seelig (1974) determined the order parameters for dipalmitoylphosphatidylcholine (DPPC) labeled at various positions along the acyl chain. These data are shown in Fig. 17 together with order parameters for

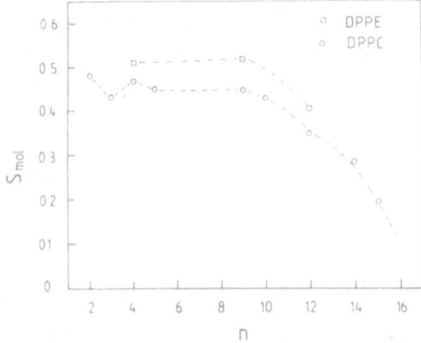

Figure 17   Order parameter S_{mol} for DPPC and DPPE at temperatures just above the phase transition temperature of DPPC (41°C) and DPPE (64°C), respectively (Seelig and Seelig, 1974; Blume *et al.*, 1982a).

DPPE labeled at three different positions (Blume *et al.*, 1982). Note, that despite the fact that the transition temperature of DPPE is 24°C higher than that of DPPC the order parameters for DPPE in the liquid-crystalline phase at 68°C are still higher than those for DPPC at approximately 45°C.

The interpretation of these order parameters S_{mol} in terms of a model for the molecular motions occurring in the liquid-crystalline phase is not unambiguous. Fast *trans-gauche* isomerization of the acyl chains together with fast long axis rotational diffusion could account for the observed splittings. However, it seems that an additional motional averaging occurs through a wobbling motion of the long axis of the molecule (Meier *et al.*, 1983; Petersen and Chan, 1977; Meier *et al.*, 1986). This additional motion can be incorporated into the order parameter concept by the assumption that the total order parameter determined from the spectra is the product of an order parameter of the whole chain S_{zz}, describing the wobbling motion of the whole molecule, and the order parameter of a specific segment, describing the *trans-gauche* isomerization process (Meier *et al.*, 1983; Petersen and Chan, 1977; Meier *et al.*, 1986; Jähnig *et al.*, 1982). With this approach the chain order parameter S_{zz} can be estimated by a line-shape analysis. For DMPC just above T_m S_{zz} is ca. 0.57 and decreases even further with increasing temperature (Meier *et al.*, 1983; 1986).

As indicated by the differences in the order parameters observed for DPPC and for DPPE as shown in Fig. 17, the order of liquid-crystalline bilayers is a function of the chemical nature of the lipid, in this case the head group. But there is also a chain length dependence. Phosphatidylethanolamines generally display higher-order parameters than the corresponding phosphatidylcholines when a certain position, namely, the 4-position of the *sn*-2 chain, is monitored. As mentioned above, the strong head group interactions via hydrogen bonds between the ethanolamine groups in PEs are probably responsible for this effect. Table I shows the residual quadrupole splittings of different phosphatidylcholines and

Table I. Splittings $\Delta\nu_Q$ in kHz of ^{2}H NMR Powder Patterns of Phospholipids in the Liquid-Crystalline Phase at Various Temperatures above $T_m{}^a$.

Phospholipid	$T = T_m + 5°C$	$T = T_m + 10°C$	75°C
DMPC	29.0	27.6	20.4
DPPC	27.4	26.3	21.1
DSPC	24.7	23.9	21.5
DMPE	31.8	30.4	28.5
DPPE	31.5	29.6	29.0
DSPE	28.7	27.8	29.1

^{a} Blume *et al.* (1982a,b); Hübner and Blume (1987).

phosphatidylethanolamines all labeled at the 4-position of the *sn*-2 chain (Hübner and Blume, 1987). When the PCs and PEs with identical chains are compared at the same reduced temperature, for instance 5 or 10°C above their respective transition temperatures, the differences in the quadrupole splittings are 3–4 kHz. These differences are much larger when the lipids are compared at the same absolute temperature, for instance 75°C. In this case the differences in the splittings between the two lipid classes increase to ca. 8 kHz. So phosphatidylethanolamine bilayers are substantially more ordered than phosphatidylcholine bilayers. As mentioned above, there exists also a chain length dependence, which is small, however, when the splittings are compared at the same absolute temperature (see the values at 75°C). On the reduced temperature scale these differences increase. Interestingly the longer-chain lipids are less ordered just above their transition temperature than their shorter counterparts and this effect reverses when the splittings are compared at the same absolute temperature. The phosphatidylcholines display a stronger chain length dependence of the splittings than the phosphatidylethanolamines. Again, this is an indication of the existence of additional, i.e., strong head group interactions, in the case of phosphatidylethanolamines.

Besides labeling the different segments of the acyl chains certain positions in the head group can also be labeled, for instance the two methylene segments of the ethanolamine or choline residue, or the three methyl groups attached to the nitrogen in the choline (Seelig and Gally, 1976; Seelig *et al.*, 1977; Blume *et al.*, 1982a; Gally *et al.*, 1975). Labeling specific carbon atoms in the glycerol backbone is more complicated but has also been accomplished (Blume *et al.*, 1982a; Gally *et al.*, 1981; Strenk *et al.*, 1985). The ^{2}H spectra of lipids labeled in the choline or ethanolamine head group all display very narrow powder patterns with splittings between 2 and 10 kHz depending on the particular labeled position, the methyl labeled PCs giving the smallest and the PEs labeled at the carbon atom next to the phosphate giving the largest splitting. These narrow powder patterns indicate that besides rotational diffusion and wobbling of the whole molecules additional motional averaging processes are present. Isomerization around several single bonds is possible and the head groups can assume a large number of different conformations. However, the average orientation of the head group remains parallel to the bilayer surface (Gally, et al., 1975).

Spectra of phospholipids labeled in the glycerol part of the molecule display characteristic splitting depending on the position in the glycerol backbone. In addition, the two deuterons at the 1- and 3-positions give different powder patterns, an indication that they are motionally and/or conformationally inequivalent (Gally et al., 1981; Strenk et al., 1985). These observations can be interpreted in different ways. It could be that isomerization takes place around the C_1-C_2 bond of the glycerol with different populations for the *trans* and *gauche* conformers. A similar asymmetric isomerization could also occur at the C_2-C_3 bond, as these processes have been observed for lipids in solution and in micelles of short-chain phospholipids (Hauser et al., 1981; Meulendijks et al., 1986). An alternative explanation is a relatively rigid glycerol backbone conformation but with director fluctuations leading to the observed splittings (Strenk et al., 1985). A definite decision on the type of motion present in the glycerol backbone part of the lipid molecules cannot be made at the present time. It should be mentioned that motional inequivalence of two deuterons in the same segment is not restricted to the glycerol part, but was observed first for the 2-position of the sn-2 chain in phospholipids (Seelig and Seelig, 1975).

3.4. Phospholipids in the Gel Phase

While ^2H NMR spectra of lipids in the liquid-crystalline state are relatively easy to record because of their narrow splittings—even a one-pulse experiment with Fourier transformation of the free induction decay will give a spectrum from which the splitting $\Delta\nu_{Q_\perp}$ can be determined—spectra of lipids in the gel state are more difficult to obtain because of their great width. The application of the quadrupole echo sequence (Davis et al., 1976; Solomon, 1958) and the availability of radiofrequency transmitters with high enough power to obtain short (1-2 μsec) 90° pulses (Griffin, 1981) made it possible to record good quality deuterium powder patterns of lipids in the rigid limit and the intermediate exchange regime. With the appropriate line shape simulation programs it was then possible to extract detailed information on the type and dynamics of motions in lipid bilayers (Spiess and Sillescu, 1981; Blume et al., 1982a; Meier et al., 1983, 1986; Huang et al., 1980).

One of the first studies of phospholipids in the gel state was an investigation of chain perdeuterated DPPC (Davis, 1979). The use of perdeuterated chains has the obvious advantage of giving a better signal-to-noise ratio. The disadvantage, of course, is the loss of information on the motional behavior of specific segments of the chains, as all resonances from the methylene segments overlap. An important result of this study was the conclusion, obtained from a second moment analysis (Slichter, 1978; Davis, 1979), that even 15–20°C below the phase transition of DPPC the lipid molecules are still rapidly rotating around their long axis, and that there is still some hydrocarbon chain disorder by *trans–gauche* isomerization. Only when the bilayers are cooled down to 0°C does the long axis rotation slow down, and only then it is in the intermediate exchange regime on the deuterium time scale.

For a single CD_2 group in a aliphatic chain the spectrum for the case of axial diffusion around the long axis of the molecule can be calculated for different rotation rates. The simulation procedure (Wittebort *et al.*, 1982; Blume *et al.*, 1982a; Huang *et al.*, 1980) is based on the chemical exchange model described by Abragam for a two-site exchange (Abragam, 1961) and includes corrections for power roll-off (Bloom *et al.*, 1980) and line shape changes due to the application of the quadrupole echo sequence (Spiess and Sillescu, 1981). The rotational diffusion is approximated in this model by a three- or sixfold rotational jump. This model, however simple it is, seems appropriate as the aliphatic chains are ordered in a hexagonal or quasihexagonal lattice and rotation probably proceeds via rotational jumps into empty lattice sites. Figure 18 shows simulated spectra for a C–D vector executing three- or sixfold rotational jumps with the bond direction perpendicular to the rotation axis, i.e., the case of an all-*trans* aliphatic chain. The rotational rate constant decreases from top to bottom. Particularly in the intermediate exchange regime characteristic changes of the line shape are observed. In the fast limit the width of the powder pattern is reduced by a factor of 2 as $\theta = 90°$ [see equation (26)]. So for lipids labeled in the aliphatic chains spectra like those in Fig. 18 should be observed, when the chains are in an all-*trans* conformation and the only motion is axial rotation.

Figure 19 shows deuterium spectra of $2(4,4-^2H_2)DPPE$ at a whole range of different temperatures (Blume *et al.*, 1982a). Again, DPPE was chosen because its thermotropic behavior is particularly simple. It displays only the main transition from the gel to the liquid-crystalline phase and its chains are perpendicular to the bilayer surface in the gel phase (McIntosh, 1980).

Above T_m a sharp powder pattern is observed as described in the previous section. Cooling the system below the phase transition leads to drastic changes in the line shape, as the width of the spectrum increases dramatically. If the molecules were executing only rotational jumps in the fast limit, spectra like those at the top of Fig. 18 should be observed.

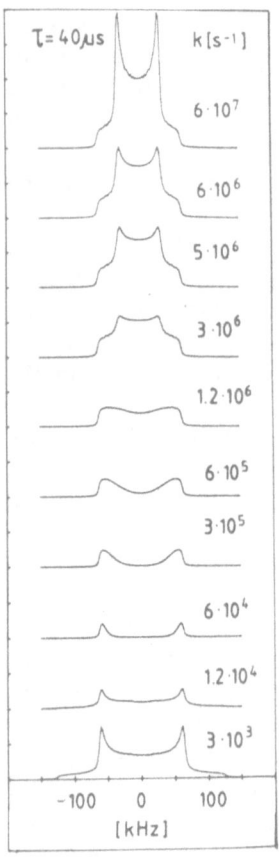

Figure 18. Line-shape simulations for a threefold rotational jump of a C–D vector with the bond direction perpendicular to the rotation axis. τ is the pulse spacing in the quadrupole echo sequence and k is the jump rate constant.

Comparison with the experimental spectra in Fig. 19 shows that the width of the powder patterns is reduced by more than a factor of 2 compared to the rigid lattice powder pattern and that the spectra do not display the characteristic sharp perpendicular edges of fast limit spectra. Particularly at lower temperatures the spectra become broad and featureless, and just above 0°C almost flat-topped spectra are observed. These line shapes clearly indicate that rotational diffusion is slowing down and approaching the intermediate exchange regime (see Fig. 18). However, the spectra also indicate that not only rotational motion can be present, but that there has to be some additional type of motional averaging present that accounts for the rounding of the edges of the spectra. These line shapes have been analyzed by computer simulations using the following model (Blume *et al.*, 1982a). The molecules are allowed to execute threefold rotational jumps around the long axis of the molecule as described before. In addition to

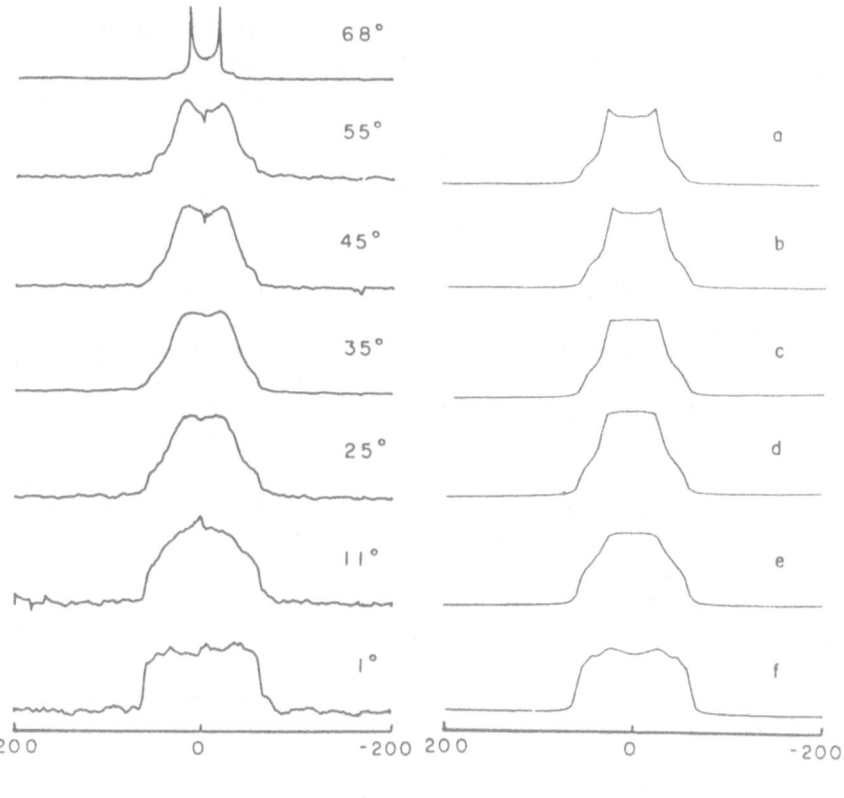

Figure 19. Left: ^{2}H NMR spectra of 2(4,4-^{2}H$_2$)DPPE in 50 wt% deuterium depleted water (Blume *et al.*, 1982a). Right: Line-shape simulations with R_R = rotational jump rate constant, R_{tg} = *trans–gauche* isomerization rate constant, P_g = *gauche* probability. (a) R_R = 4.7×10^6 sec^{-1}, R_{tg} = 5.3×10^5 sec^{-1}, P_g = 0.07; (b) R_R = 3.9×10^6 sec^{-1}, R_{tg} = 4.6×10^5 sec^{-1}, P_g = 0.06; (c) R_R = 3.1×10^6 sec^{-1}, R_{tg} = 3.4×10^5 sec^{-1}, P_g = 0.06; (d) R_R = 2.8×10^6 sec^{-1}, R_{tg} = 2.5×10^5 sec^{-1}, P_g = 0.05; (e) R_R = 1.9×10^6 sec^{-1}, R_{tg} = 1.9×10^5 sec^{-1}, P_g = 0.04; (f) R_R = 1.3×10^6 sec^{-1}, R_{tg} = 1.7×10^5 s^{-1}, P_g = 0.02.

this motion *trans–gauche* isomerization is allowed, i.e., a C–D vector can jump from an orientation perpendicular to the rotation axis to a *gauche* site where the angle of the C–D bond is 35.3° with respect to the long axis. This leads to overall six different sites, three *trans* and three *gauche* sites. The probabilities for the *gauche* populations are adjustable parameters, as are the rate constants for the rotational jumps and the *trans–gauche* isomerization. While this model is certainly a simplification of the real situation, it reproduces the experimental spectra quite well (see the right-hand side of Fig. 19).

As mentioned above, the parameters used in the simulations are the probability for a *gauche* conformation P_g, the rate constant for *trans-gauche* isomerization R_{tg}, and the rate constant for the rotational threefold jumps, R_R. A rough estimate for the *gauche* probability P_g can be obtained from the splitting of the parallel edges $\overline{\Delta\nu_{Q_\parallel}}$ of the spectra. $\overline{\Delta\nu_{Q_\parallel}}$ is ca. 115 kHz for the 25°C spectrum. This is less than the expected 125 kHz for an all-*trans* chain. So *trans-gauche* isomerization is occurring at room temperature. P_g can be calculated to be ca. 0.05, i.e., there is a 5% probability for a *gauche* conformation at this particular site. The rate constant for the rotational jumps decreases from ca. $5 \times 10^6\, \text{sec}^{-1}$ to $1 \times 10^6\, \text{sec}^{-1}$ at 1°C, where the intermediate exchange regime is approached. The deuterium spectra are thus in accordance with the results of the ^{13}C spectra (see above), namely, that rotational motion around the long axis of the molecule is fast on the ^{13}C time scale, i.e., has correlation times shorter than 1×10^{-5} sec.

The spectra shown in Fig. 19 are not plotted scaled to the absolute intensity of the quadrupole echo as was done for the simulations shown in Fig. 18. In the intermediate exchange regime a characteristic decrease in spectral intensity is observed, which is particularly pronounced in the middle part of the spectrum (see Fig. 18). This is also observed experimentally. Figure 20 shows three spectra of $2(8,8\text{-}^2\text{H}_2)$DPPE at different temperatures and the corresponding simulations, in this case all spectra plotted on an absolute intensity scale. This figure shows that the model used in the simulation procedure not only reproduces the line shapes but also the absolute spectral intensities quite well. The top spectrum in this figure shows a spectrum at 63°C, a temperature inside the phase transition region of DPPE (see Fig. 9). Here the intensity is also reduced, but not because the rotational motion is in the intermediate exchange regime. At this temperature the molecules are exchanging by lateral diffusion between gel and liquid-crystalline phase domains, and these exchange processes happen to be in the intermediate exchange regime thus decreasing the echo intensity.

^{2}H NMR spectra of DPPE labeled at positions further down the chain look more or less identical to the spectra shown in Fig. 19 (see also Fig. 20) (Blume *et al.*, 1982a). The slight differences are in the rotational jump rates; they increase for positions 8 and 12, and in the *gauche* probabilities, which are somewhat larger for the 12-position compared to the 8- and 4-positions. The increase in rate constant for the rotational jumps can be rationalized by the assumption of additional averaging mechanisms like torsional oscillations (Blume *et al.*, 1982a; Cameron *et al.*, 1981). The increase in *gauche* probability for methylene groups near the chain ends is comparable to the decrease in order parameter observed in the liquid-crystalline phase (see Fig. 17). So in the gel phase there exists a similar order parameter profile as in the liquid-crystalline phase (Blume *et al.*, 1982a; Meier *et al.*, 1986).

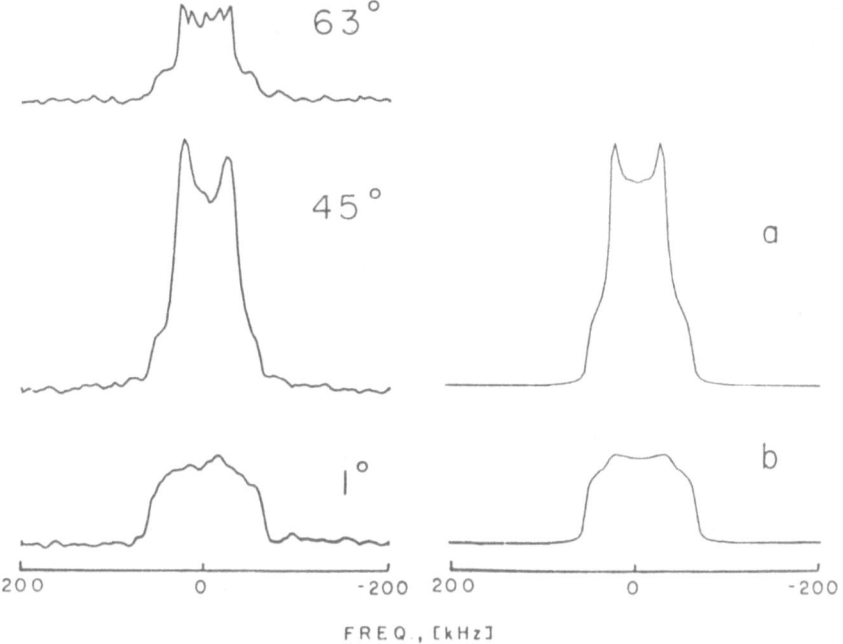

Figure 20. Left: ^{2}H NMR spectra of 2(8,8-^{2}H$_2$)DPPE in 50 wt% deuterium depleted water plotted scaled to the absolute intensity of the quadrupole echo (Blume *et al.*, 1982). Right: Line-shape simulations showing the decrease in echo intensity when intermediate exchange rates are encountered. (a) $R_R = 5.3 \times 10^6\,\text{sec}^{-1}$, $R_{tg} = 9.4 \times 10^5\,\text{sec}^{-1}$, $P_g = 0.05$; (b) $R_R = 1.7 \times 10^6\,\text{sec}^{-1}$, $R_{tg} = 3.1 \times 10^5\,\text{sec}^{-1}$, $P_g = 0.02$.

When segments in the glycerol backbone are labeled, the motion of the whole molecule should be monitored as this should be the most rigid part of the molecule when it is in the gel phase. Figure 21 shows spectra of DPPE labeled at the C_2 position of the glycerol, so only one deuteron is present. In the gel phase the spectra are axially symmetric, indicating rotational motion, but the splittings are much larger than expected for a C–D vector perpendicular to the rotation axis. The splittings are between 95 and 110 kHz. This indicates that the C–D bond must have an orientation between 20° and 35° relative to the long axis. This requires a *gauche* bond around the C_1–C_2 bond of the glycerol, a conformation that has been shown by potential energy calculations to be one of the two low-energy conformations of lipids, and which in fact has recently been determined by X-ray diffraction as the conformation of crystalline dimethyl-PE (McAlister *et al.*, 1973; Pascher and Sundell, 1986). The simulations on the right-hand side of Fig. 21 were performed with a three-site jump model for the rotational motion and an additional two-site librational motion where the C–D vector can jump between two orientations, namely, 15° and 35° relative to the

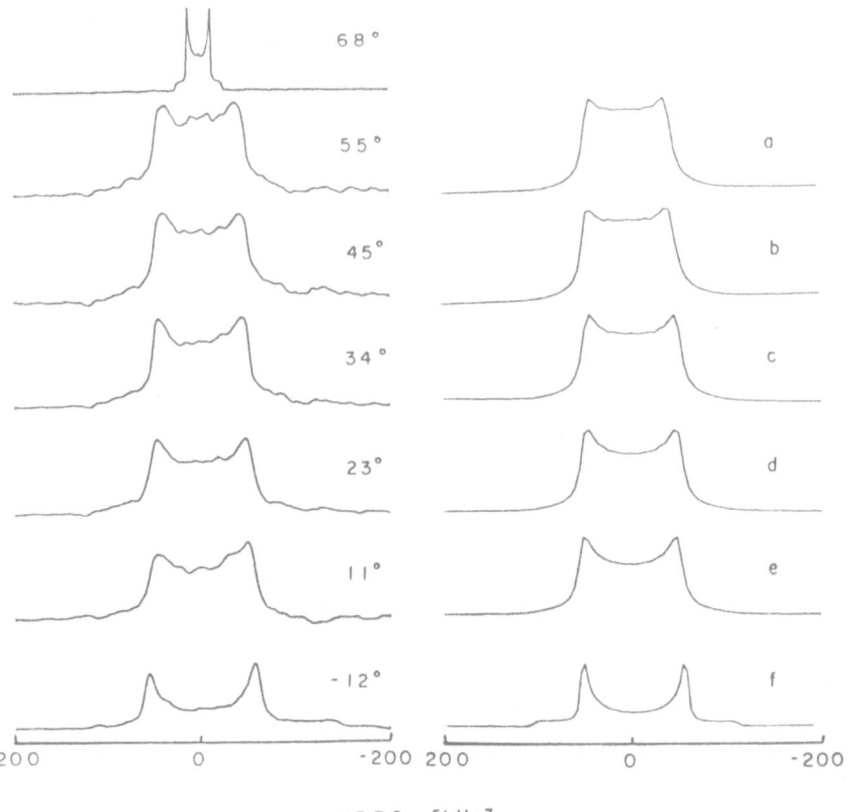

FREQ., [kHz]

Figure 21. Left: ^2H NMR spectra of (glycerol-C_2-$^2H_1$)DPPE in 50 wt% deuterium depleted water (Blume et al., 1982a). Right: Line-shape simulations using a threefold rotational jump motion (R_R) and a librational motion of the C–D bond between 15° and 35° relative to the rotation axis. (a) $R_R = 4.7 \times 10^6 \, \text{sec}^{-1}$, (b) $R_R = 3.9 \times 10^6 \, \text{sec}^{-1}$, (c) $R_R = 2.8 \times 10^6 \, \text{sec}^{-1}$, (d) $R_R = 2.4 \times 10^6 \, \text{sec}^{-1}$, (e) $R_R = 1.8 \times 10^6 \, \text{sec}^{-1}$, (f) $R_R = 9.1 \times 10^5 \, \text{sec}^{-1}$.

rotation axis. As can be seen from the data in Fig. 21 the rotational jump rates determined from these spectra are very similar to those obtained from the simulations of 2(4,4-$^2H_2$)DPPE (see Fig. 19). Thus labeling DPPE in the glycerol backbone or in the chains at a position close to the ester bonds produces spectra that contain mainly information on the motion of the whole molecule.

The lipid head group, i.e., in the case of DPPE the ethanolamine group, should be the most mobile part of the molecule, even in the gel phase. The deuterium spectra show that this is indeed the case. Figure 22 shows spectra of DPPE labeled at the CH_2 group of ethanolamine next to the phosphate. The spectral width is considerably reduced compared to the chain labeled

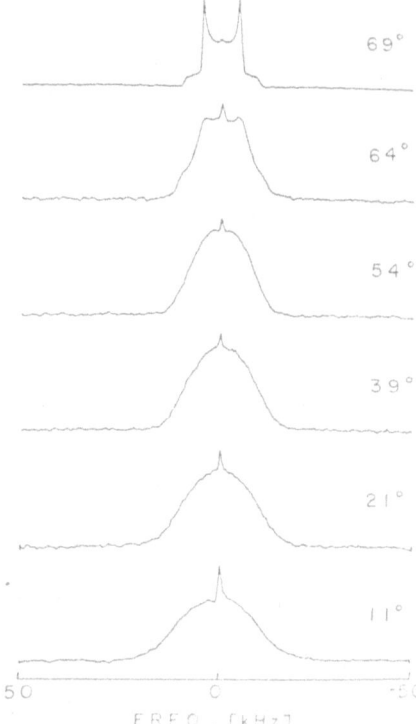

Figure 22. ²H NMR spectra of head group labeled DPPE ($-CD_2-CH_2-NH_3$) in 50 wt% deuterium depleted water (Blume *et al.*, 1982a).

lipids. The line shape is very uncharacteristic as no sharp perpendicular edges are observed but only a broad line with a half-height width of ca. 15–20 kHz. This type of line can only be qualitatively interpreted. Apparently a whole range of different conformations is accessible by the ethanolamine group and these are rapidly interconverting. At temperatures just above 0°C the intensity decrease is much larger than observed for the chain-labeled compounds. This is to be expected when the motions are almost isotropic (Spiess and Sillescu, 1981). This strong intensity decrease thus indicates that the head group motions are slowing down at this temperature, but that the angular fluctuations of the ethanolamine group remain large.

The analysis of the deuterium spectra of DPPE labeled at various positions in the molecule gives a fairly complete picture on the motional characteristics of this particular phospholipid in the gel as well as in the liquid-crystalline phase. This method can, of course, also be applied to other lipids. Particularly, phosphatidylcholines and cerebrosides have been analyzed in detail (Meier *et al.*, 1983, 1986; Seelig and Seelig, 1975; Ruocco *et al.*, 1985a,b). Figure 23 shows deuterium spectra of 2(4,4-²H₂)DPPC at different temperatures (Blume *et al.*, 1982b). At 1°C the spectrum is very similar to the line shape observed for DPPE at 11°C.

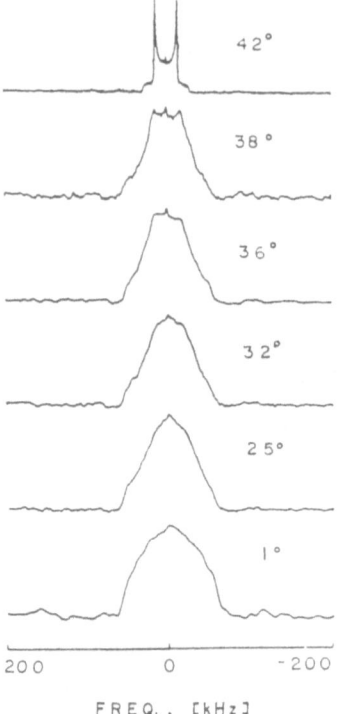

Figure 23. ^{2}H NMR spectra of 2(4,4-^{2}H$_2$)DPPC in 50 wt% deuterium depleted water. In the region between the pre- (35°C) and the main transition (41°C) two-component spectra are observed. This is particularly evident for the spectrum taken at 38°C (see also Fig. 11) (Blume *et al.*, 1982b).

A closer inspection reveals that the total width, i.e., the separation of the parallel edges of the spectrum, is larger for DPPE compared to DPPC. This indicates a slightly looser packing of the DPPC molecules. In the ^{13}C NMR experiments (see Section 3.1), we observed two-component spectra in the temperature range between 28 and 40°C, with one component growing at the expense of the other. In the ^{2}H spectra this is not so obvious, except maybe for the 38°C spectrum. Here the separation of the perpendicular edges of the spectrum $\overline{\Delta\nu_{Q_\perp}}$ is smaller than the separation of the parallel edges $\overline{\Delta\nu_{Q_\parallel}}$, indicating that more than one component may be present. Further evidence comes from the absolute intensities of the spectra in this temperature range. They decrease just below the main phase transition and then increase again at 25°C. This intensity decrease is again due to exchange between the two components via lateral diffusion in the rippled phase below the main transition. The exchange rates could be more easily calculated from the ^{13}C spectra shown in Fig. 11. They are in the 500–5000-Hz range (Wittebort *et al.*, 1982) and are thus at the lower end of the intermediate exchange regime of the deuterium time scale. As the width of the deuterium powder pattern for the two components changes by ca. 20 kHz these exchange rates can still affect the intensity of the quadrupole echo.

²H NMR spectra of chain-labeled galactosylcerebrosides have been recorded by Huang et al. (1980). These spectra show no indication of long axis rotation in agreement with the ¹³C spectra of the same lipids (see Fig. 12). The ²H spectra of galactosylcerebrosides can be simulated by assuming two- or three-site *trans–gauche* isomerization with increasing rates and probabilities for the *gauche* site at higher temperature. Just below T_m the spectra are similar to $\eta = 1$ line shapes, i.e., the averaged tensor is nearly totally asymmetric [see equation (20)]. In methylene chains this occurs when *gauche* and *trans* probabilities are nearly equal. As the angle between the *gauche* and *trans* site is 109°, the resulting line shape in the fast limit is identical to the one observed for a 71° jump, as jumps of θ and $180° - \theta$ give identical results.

An example for the existence of long-axis rotation down to very low temperatures is the case of the ether lipid dihexadecylphosphatidylcholine. This lipid forms an interdigitated gel phase with a thickness of about one length of a molecule (Ruocco et al., 1985a,b). The ²H NMR spectra indicate that in this phase long axis rotation can persist down to very low temperatures, even below the freezing point of water. This is also the case for a mixed system, namely, an equimolar mixture of 1-palmitoyl-lyso-phosphatidylcholine with palmitic acid. Thermodynamically this mixture behaves very similarly to the parent compound dipalmitoyl-phosphatidylcholine (Allegrini et al., 1983). Below the phase transition fast limit powder patterns are observed when the palmitic acid is labeled, indicating fast axial diffusion. This allows a more precise determination of order parameters. It could be shown that the *gauche* probability is low in the middle part of the palmitic acid chain (ca. 5%) but that it increases towards the chain end (Allegrini et al., 1983). This is in good agreement with the results obtained from computer simulations of chain-labeled DPPE in the gel phase (see Fig. 19).

3.5. Binary Mixtures of Phospholipids

As described before for the ¹³C spectra the possibility of labeling one component of a mixture at a time is of great advantage for the analysis of lipid mixtures. This applies also to ²H NMR. The analysis of the spectra will yield information on the miscibility of different lipids via changes in their motional behavior. The example for a phospholipid mixture will be the same as shown above in the ¹³C NMR section, namely, the DPPC–DPPE mixture. The ¹³C spectra presented in Fig. 13 indicated domain formation with negligible exchange in the temperature range of the phase transition. These two-component line shapes can also be observed in the ²H NMR spectra. Figure 24 gives as an example spectra of an equimolar DPPC–DPPE mixture. On the left-hand side DPPC is the labeled lipid, and on the right DPPE. The low-temperature spectra of the two lipids look quite similar and there is also not much difference to the spectra of the pure compounds. Thus the order and dynamics of the molecules seem to be very similar. This had also been inferred from the results of the ^{13}C NMR experiments (see Fig. 13). In the temperature range of the phase transition two types of powder patterns are superimposed, one with reduced splitting and sharp perpendicular edges characteristic for liquid-crystalline lipid, and the other similar to the gel phase spectrum. As the exchange rates were slow on the ^{13}C NMR time scale they should also not affect the deuterium spectra. However, in the two-phase region the quadrupole echo intensity decreases, which indicates that at least some of the molecules, presumably those near a domain boundary, must exchange with rates comparable to the differences in the quadrupole splittings of the two components (Blume *et al.*, 1982b). As mentioned above, the line shapes for the two components are very similar at temperatures below the phase transition region. Nevertheless a close inspection reveals that there are slight differences in total width of the

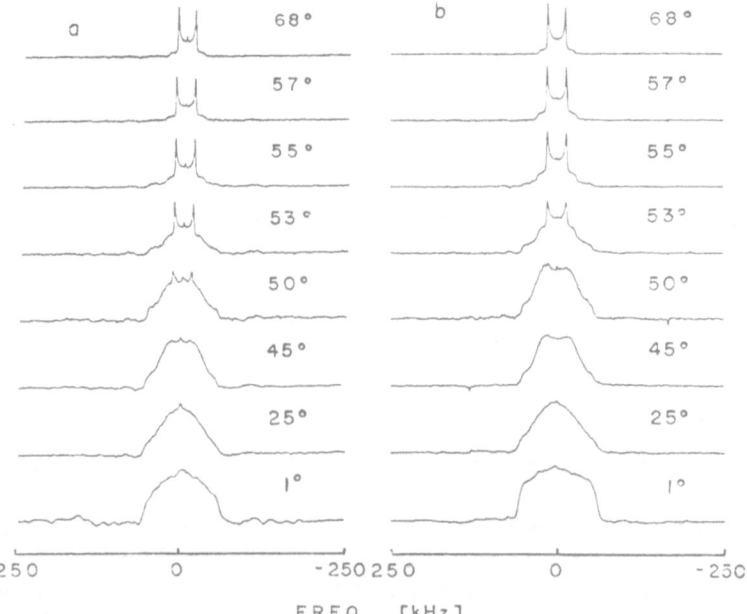

FREQ. , [kHz]

Figure 24. ^{2}H NMR spectra of 1 : 1 mixtures of (a) 2(4,4-^{2}H$_2$)DPPC–DPPE and (b) DPPC–2(4,4-^{2}H$_2$)DPPE in 50 wt% deuterium depleted water. In the phase transition region between 45 and 57°C a superposition of liquid-crystalline-like powder patterns with broad lines characteristic for the gel phase are observed (Blume *et al.*, 1982b).

spectra. The powder patterns of DPPE are always ca. 5 kHz wider than those of DPPC. So some of the characteristics of the pure components are retained in the mixture. This could arise from the fact that DPPE is still able to form a hydrogen bond to the next molecule, even when it is DPPC and not DPPE, while DPPC with its choline head group cannot form hydrogen bonds at all. The average width of the powder pattern shows a linear dependence on composition, so almost ideal mixing behavior in the gel phase is observed, in agreement with the ¹³C NMR results (see Fig. 14).

A similar analysis of the quadrupole splittings in the liquid-crystalline phase also shows slight differences in the order and/or dynamics of the two components (see Fig. 25). This plot of the splittings versus the composition shows small differences between the two components in the mixture, probably for the same reasons as discussed above for the gel phase spectra. A plot of the average splitting (see Fig. 25) versus the concentration indicates negative deviations from ideal mixing behavior. This means that the mixed

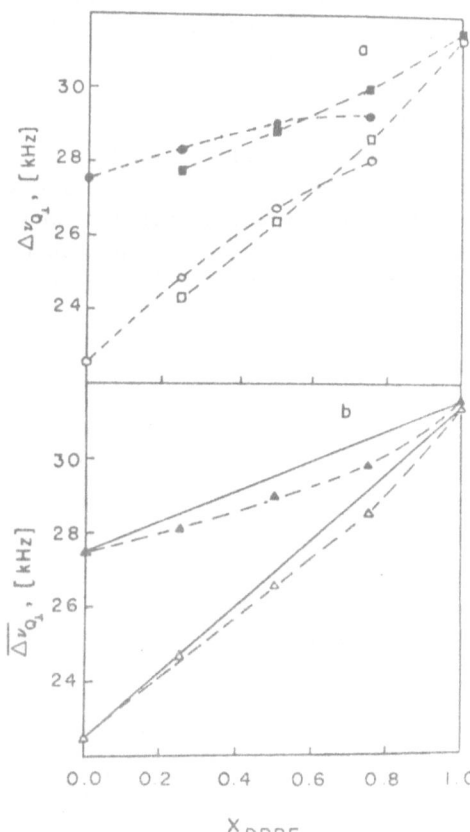

Figure 25. (a) Quadrupole splittings $\Delta\nu_{Q_\perp}$ in the liquid-crystalline phase of DPPC-DPPE mixtures: O, $2(4,4$-$^2H_2)$DPPC at 68°C; ●, $2(4,4$-$^2H_2)$DPPC at $T = T_{comp}$; □, $2(4,4$-$^2H_2)$DPPE at 68°C; ■, $2(4,4$-$^2H_2)$DPPE at $T = T_{comp}$. (b) Average quadrupole splitting $\overline{\Delta\nu_{Q_\perp}}$ calculated from the values in Fig. 25a at △, 68°C and ▲, $T = T_{comp}$ (Blume *et al.*, 1982b).

liquid-crystalline phase is a little more "fluid" or expanded than expected
from the characteristics of the pure components. These deviations from
ideal behavior are relatively subtle, as is to be expected when the two
components differ only in their head group structure and not in their chain
length. The deviations should be more pronounced for mixtures of two
components with different chain lengths. Figures 26 and 27 show as examples
the splittings of the powder patterns observed for mixtures of DMPE with
DSPC and for DMPC with DSPE, i.e., binary mixtures where the lengths
of the acyl chains differ by four CH_2 groups (Hübner and Blume, 1987).
In both cases much stronger deviations from ideality can be seen, the
nonideality being greater for the DMPC–DSPE mixture. This is because of
the large difference in transition temperatures of the two components, which
are 24°C for DMPC and 72°C for DSPE. When the chain lengths are reversed
the T_m difference decreases to 5°C. Equimolar mixtures of DMPE with
DSPC show a peculiar behavior as the deviations from ideality are almost
zero for this composition (see Fig. 26). It could be possible that a 1:1
mixture forms a particularly stable liquid-crystalline phase or that even 1:1
complex formation occurs. A particularly strong deviation from ideal
behavior has also been observed by DSC for this mixture, as this has an
extraordinarily high transition enthalpy (Blume and Ackermann, 1974). A
second possibility that can be discussed is the existence of a eutectic point
as suggested by Sugar and Monticelli (1983). NMR experiments cannot
distinguish between these two possibilities, but the results clearly show that
the order and motional behavior of lipids in the liquid-crystalline phase
can be quite varied, and that mixtures of two phospholipids behave
differently than expected from a simple averaging of the properties of the

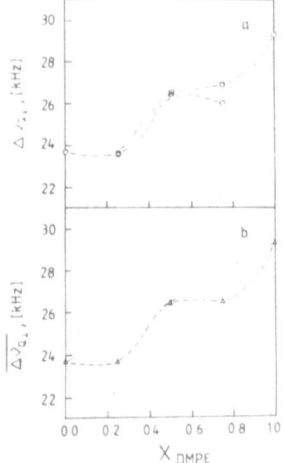

Figure 26. (a) Quadrupole splitting $\Delta\nu_{Q\perp}$ in the liquid-
crystalline phase of DMPE–DSPC mixtures at 60°C (Hübner
and Blume, 1987). ◯, 2(4,4-$^2H_2$)DSPC; ☐, 2(4,4-$^2H_2$)DMPE.
(b) Average quadrupole splitting at 60°C calculated from data
in Fig. 26a.

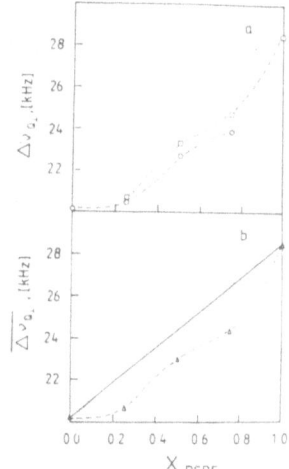

Figure 27. (a) Quadrupole splitting $\Delta\nu_{Q_-}$ in the liquid-crystalline phase of DMPC-DSPE mixtures at 75°C (Hübner and Blume, 1987). ○, $2(4,4\text{-}^2H_2)$DMPC; □, $2(4,4\text{-}^2H_2)$DSPE; (b) Average quadrupole splitting at 75°C calculated from data in Fig. 27a.

two components. For the three cases of phosphatidylcholine-phosphatidylethanolamine mixtures shown, one could see in every mixture a negative deviation from ideality. The quadrupole splittings of lipids in the mixture were reduced, i.e., the order parameters were smaller than expected. Whether these differences are due to reduced conformational order in the chains or a reduction of the wobbling motion of the lipid molecules cannot be decided at the present stage.

3.6. DPPE–Cholesterol Mixtures

The ¹³C spectra of DPPE–cholesterol mixtures (see Fig. 15) indicated exchange between the two components with different conformation with exchange rates between 1 and 10 kHz. In the ²H NMR spectra of chain-labeled DPPE mixed with cholesterol two components are not directly visible; rather the line shapes seem to indicate only one component (see Fig. 28). With increasing cholesterol concentration, sharp, well-resolved fast limit powder patterns appear at progressively lower temperature. Cholesterol is obviously acting as a spacer preventing cooperative interactions between the chains and polar interactions via hydrogen bonds. The quasihexagonal lattice is broken up and the molecules are now able to rotate almost freely. The spacer effect, i.e., the breaking-up of the head group interactions of DPPE, can be nicely seen when spectra of head group labeled DPPE are recorded. Figure 29 shows the splitting versus cholesterol concentration of ethanolamine-labeled DPPE. Clearly a decrease in splitting is observed, indicating more motional freedom for the head group when cholesterol is intercalated between the DPPE head groups. The distances

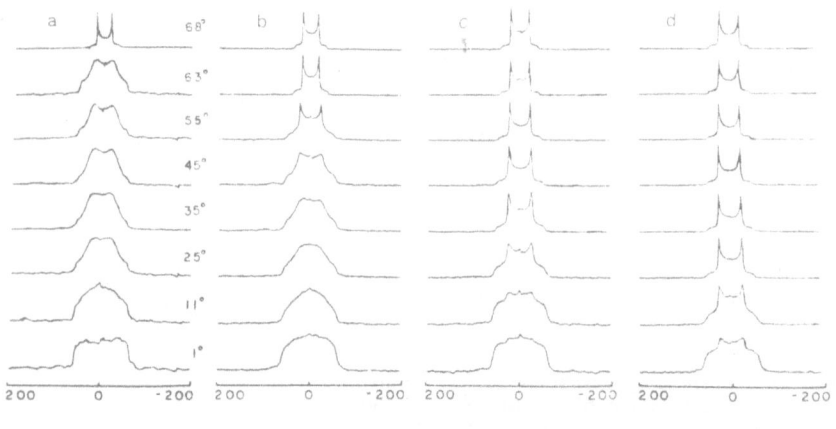

FREQUENCY, [kHz]

Figure 28. ^{2}H NMR spectra of 2(4,4-^{2}H$_2$)DPPE in mixtures with cholesterol at different temperatures; (a) 0; (b) 10; (c) 30; (d) 50 mol% cholesterol, respectively (Blume and Griffin, 1982).

between the ethanolamine NH$_3$ and the next phosphate are now too large to form a strong hydrogen bond.

For the chain-labeled DPPE the quadrupole splittings are larger in mixtures with cholesterol compared to the pure liquid-crystalline phase. This is illustrated in Fig. 30, where $\overline{\Delta \nu_Q}$ is plotted at constant temperature versus cholesterol content. Particularly from the 68° curve it is evident that *trans–gauche* isomerization and possibly chain fluctuations are reduced by the addition of cholesterol. From the right-hand side of Fig. 30 one can see that samples with high cholesterol content show a more or less continuous decrease of the splitting with temperature. Mixtures with 10–30 mol% cholesterol, however, still undergo a cooperative sigmoidal change of their splittings. If these curves at low cholesterol content would be normalized they would more or less coincide with the normalized DSC transition curves shown in Fig. 16. Thus the ^{2}H NMR spectra can be partly correlated with thermotropic properties as measured by DSC. However, it is obvious that superimposed on the sigmoidal change of the splitting there is a continuous change of the properties of the system that can only be detected by NMR methods and not by DSC.

Though the ^{2}H spectra are remarkable in that they lack a second well-resolved component in contrast to the ^{13}C spectra shown in Fig. 15, they can nevertheless be simulated with two components exchanging with rate constants determined from the ^{13}C spectra. Figure 31 shows experimental spectra of a sample with 20 mol% cholesterol together with computer simulations. These calculations were performed with the following model:

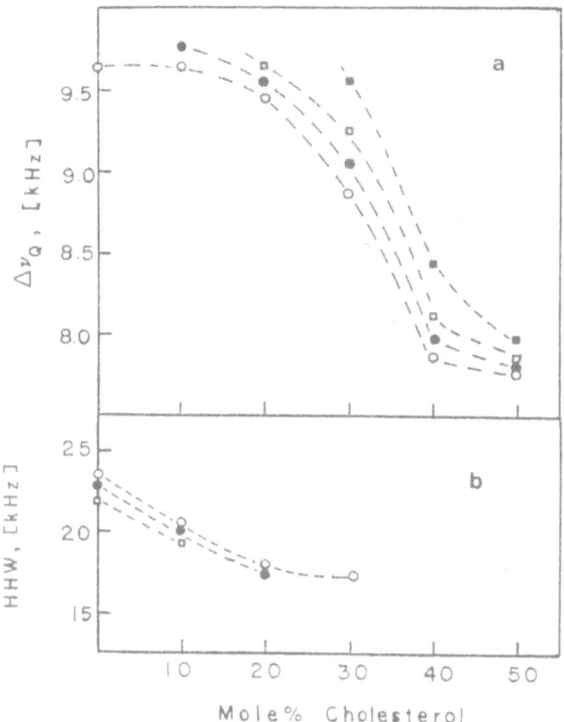

Figure 29. (a) Residual quadrupole splitting $\Delta\nu_Q$ for head group labeled DPPE in DPPE–cholesterol mixtures as a function of cholesterol content: O, 68°C; ●, 63°C; ☐, 55°C; ■, 45°C. (b) Width at half-height of the broad component at O, 1°C; ●, 11°C; ☐, 23°C (Blume and Griffin, 1982).

one component exhibits a line shape similar to pure DPPE at this particular temperature, while the second component is characterized by fast axial diffusion and increased *trans-gauche* isomerization. The fractions and exchange rates were taken from the simulations of the ¹³C spectra in Fig. 15 (Blume and Griffin, 1982). The simulations show excellent agreement with the experimental spectra. Note that this type of exchange only leads to a slight broadening of the line shape and not to the appearance of two resolved splittings. That exchange processes are present can only be seen from the reduced integral intensities of the powder patterns (the spectra and the simulations in Fig. 31 are plotted scaled to the absolute intensity of the quadrupole echo). This intensity decrease is also well reproduced by the simulations. The exchange rates can be interpreted in terms of a mean separation of the two components. As they exchange by lateral diffusion a mean square displacement $x^2 = 4D_t t$ can be calculated. An exchange rate of $2 \times 10^3 \, \mathrm{sec}^{-1}$ is equivalent to an exchange lifetime t of 5×10^{-4} sec.

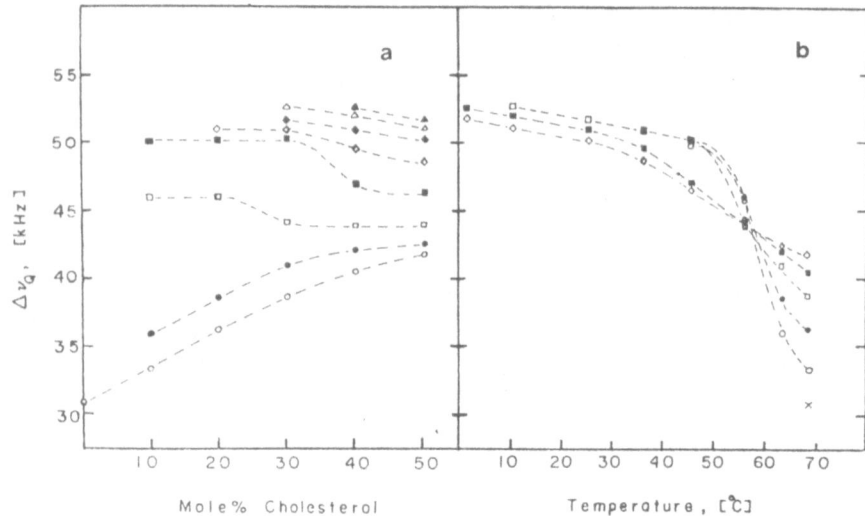

Figure 30. (a) Residual quadrupole splitting $\Delta\nu_{Q_\perp}$ for $2(4,4\text{-}^2H_2)$DPPE-cholesterol mixtures versus cholesterol content: $\bigcirc$, 68°C; $\bullet$, 63°C; $\square$, 55°C; $\blacksquare$, 45°C; $\diamond$, 35°C; $\blacklozenge$, 25°C, $\triangle$, 11°C; $\blacktriangledown$, 1°C. (b) versus temperature: $\times$, 0; $\bigcirc$, 10; $\bullet$, 20; $\square$, 30; $\blacksquare$, 40; $\diamond$, 50 mol% cholesterol, respectively (Blume and Griffin, 1982).

The lateral diffusion coefficient D_l in DMPC–cholesterol mixtures has been determined to ca. 2×10^{-9} cm^2/sec at 16°C and 25 mol% cholesterol (Rubinstein *et al.*, 1979). This would yield a value for the mean square displacement of ca. 40,000 Å^2, an area occupied by ca. 800 molecules. As the lateral diffusion coefficient in DPPE–cholesterol bilayers might be smaller, this indicates that the two components observed by NMR must be quite close together. From both types of NMR experiments the following picture emerges: Incorporation of cholesterol reduces the PE head group interactions via hydrogen bonds by a spacer effect. The distance between the ethanolamine NH$_3$ and the phosphate becomes too large to form a strong hydrogen bond. The rigid cholesterol ring system reduces the cooperative interactions between the acyl chains. At intermediate cholesterol concentrations (20–40 mol%) the cholesterol molecules are homogeneously distributed in the lipid bilayer. A considerable fraction of the PE molecules, likely those next to a cholesterol, show fast axial diffusion and increased *trans–gauche* isomerization, though to a lesser extent than pure DPPE in the liquid-crystalline phase. Other PE molecules, i.e., those farther away from the cholesterol, display similar characteristics to gel phase DPPE when the cholesterol content is still relatively low. These two types of molecules exchange by lateral diffusion with rates indicating that the two species are close together. Thus the DPPE–cholesterol system, though still displaying

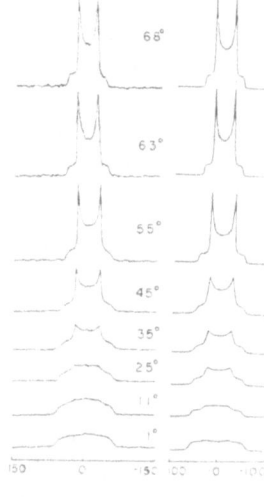

Figure 31. Left: ²H NMR spectra of a 2(4,4-²H₂)DPPE–20 mol% cholesterol mixture. Right: Line shape simulations using data derived from the simulations of the ¹³C-spectra in Fig. 15 (see also Fig. 16). The spectra are plotted scaled to the absolute intensity of the quadrupole echo (Blume and Griffin, 1982).

a calorimetric endotherm similar to that observed for binary phospholipid mixtures, does not show phase separation into larger gel and liquid-crystalline domains. It stays homogeneously mixed over the whole temperature range of the transition. These NMR results are supported by electron microscopy, X-ray diffraction (Hui and He, 1983), and by pressure jump relaxation experiments (Blume and Hillmann, 1986). The latter results indicate a change of the order of the phase transition from first order with phase separation between gel and liquid-crystalline domains at low cholesterol concentration to second order at higher cholesterol content, the exact critical concentration at which this change occurs depends on the nature of the phospholipid and on the temperature. A second-order phase transition is characterized by a discontinuous change of the heat capacity at a certain temperature, by a homogeneous distribution of the two species, and by a lack of latent heat of melting. The combination of NMR techniques with DSC, X-ray, and kinetic experiments, all intrinsic methods that do not require the introduction of a potentially perturbing label, thus allows a comprehensive characterization of the complex behavior of lipid-cholesterol mixtures.

4. SUMMARY

The combination of different NMR methods like ¹³C chemical shift anisotropy spectroscopy of specific labeled lipids with ²H quadrupole echo spectroscopy is a potent tool to determine structure, conformation, and inter- as well as intramolecular dynamics of lipids in the different mesomorphic phases. Particularly useful is the extension of these methods to lipid mixtures, which are more relevant to the problems encountered in the studies of biological membranes. As it is possible to label one particular species of the mixture at a time, we can study the motional behavior of this component without interference from the other species. Thus very detailed information on the microscopic behavior can be obtained. Thus NMR spectroscopy has a clear advantage in these cases over other methods that only monitor the average over all components of the system. The examples shown above have only dealt with pure and mixed lipid systems. These methods can of course also be used to study lipid-protein interactions and biological membranes, a topic not within the scope of this chapter.

ACKNOWLEDGMENTS. Most of the work described in this chapter was performed during the author's stay with the Solid State NMR Group of Dr. R. G. Griffin at the National Magnet Laboratory of the Massachusetts Institute of Technology, Cambridge, Massachusetts, U.S.A., to whom I am very much indebted. I also want to thank D. M. Rice, T. H. H. Huang, S. K. DasGupta, and particularly R. J. Wittebort for their help. I also thank W. Hübner, Institut für physikalische Chemie der Universität Freiburg for providing me with his ^{2}H NMR results on DMPC-DSPE and DMPE-DSPC mixtures, for support by the Deutsche Forschungsgemeinschaft through grant Nos. B1 182/2-4 and SFB 60/D-4, and by the Fonds der Chemischen Industrie.

REFERENCES

Abragam, A., 1961, *The Principles of Nuclear Magnetism*, Oxford University Press, London.
Allegrini, P. R., van Scharrenburg, G., de Haas, G. H., and Seelig, J., 1983, *Biochim. Biophys. Acta* **731**:448.
Bloembergen, N., and Rowland, T. J., 1955, *Phys. Rev.* **55**:1679.
Bloom, M., Davis, J. H., and Valic, M. I., 1980, *Can. J. Phys.* **58**:1510.
Blume, A., 1980, *Biochemistry* **19**:4908.
Blume, A., and Ackermann, Th., 1974, *FEBS Lett.* **43**:71.
Blume, A., and Griffin, R. G., 1982, *Biochemistry* **21**:6230.
Blume, A., and Hillmann, M., 1986, *Eur. Biophys. J.* **3**:343.
Blume, A., Rice, D. M., Wittebort, R. J., and Griffin, R. G., 1982a, *Biochemistry* **21**:6220.
Blume, A., Wittebort, R. J., DasGupta, S. K., and Griffin, R. G., 1982b, *Biochemistry* **21**:6243.
Cameron, D. G., Casal, H. L., Mantsch, H. H., Boulanger, Y., and Smith, I. C. P., 1981, *Biophys. J.* **35**:1.
Chen, S. C., Sturtevant, J. M., and Gaffney, B. J., 1980, *Proc. Natl. Acad. Sci. U.S.A.* **77**:5060.
Church, S. E., Griffiths, D. J., Lewis, R. N. A. H., McElhaney, R. M., and Wickman, H. H., 1986, *Biophys. J.* **49**:597.

Cornell, B. A., 1981, *Chem. Phys. Lipids* **28**:69.

Davis, J. H., 1979, *Biophys. J.* **27**:339.

Davis, J. H., Jeffrey, K. R., Bloom, M., Valic, M. I., and Higgs, T. P., 1976, *Chem. Phys. Lett.* **42**:390.

Davis, J. H., Nichol, C. P., Weeks, G., and Bloom, M., 1979a, *Biochemistry* **18**:2103.

Davis, J. H., Maraviglia, B., Weeks, G., and Godin, G. V., 1979b, *Biochim. Biophys. Acta* **550**:362.

Davis, J. H., Clare, D. M., Hodges, R. S., and Bloom, M., 1983, *Biochemistry* **22**:5298.

Demel, R. A., and de Kruyff, B., 1976, *Biochim. Biophys. Acta* **457**:109.

Devaux, P., and McConnell, H. M., 1972, *J. Am. Chem. Soc.* **94**:4475.

Doniach, S., 1979, *J. Chem. Phys.* **70**:4587.

Estep, T. N., Mountcastle, D. B., Biltonen, R. L., and Thompson, T. E., 1978, *Biochemistry* **17**:1984.

Fukushima, E., and Roeder, St. B. W., 1981, *Experimental Pulse NMR. A Nuts and Bolts Approach*, Addison-Wesley, London.

Gally, H. U., Niederberger, W., and Seelig, J., 1975, *Biochemistry* **14**:3647.

Gally, H. U., Pluschke, G., Overath, P., and Seelig, J., 1981, *Biochemistry* **20**:4223.

Gebhardt, C., Gruler, H., and Sackmann, E., 1977, *Z. Naturforsch.* **32c**:581.

Griffin, R. G., 1981, *Methods Enzymol.* **72**:108.

Hauser, H., Pascher, I., and Pearson, R. H., 1981, *Biochim. Biophys. Acta* **650**:21.

Huang, T. H., Skarjune, R. J., Wittebort, R. J., Griffin, R. G., and Oldfield, E., 1980, *J. Am. Chem. Soc.* **102**:7377.

Hübner, W., and Blume, A., 1987, *Ber. Bunsenges. Phys. Chem.* **91** (to be published).

Hui, S. W., 1981, *Biophys. J.* **34**:383.

Hui, S. W., and He, N. B., 1983, *Biochemistry* **22**:1159.

Jacobs, R. E., and Oldfield, E., 1981, *Prog. Nucl. Magn. Reson. Spectrosc.* **14**:113.

Jähnig, F., Vogel, H., and Best, L. 1982, *Biochemistry* **21**:6790.

Janiak, M. J., Small, D. M., and Shipley, G. G., 1976, *Biochemistry* **15**:4575.

Kang, S. Y., Gutowsky, H. S., Hsung, J. C., Jacobs, R., King, T. E., Rice, D., and Oldfield, E., 1979a, *Biochemistry* **18**:3257.

Kang, S. Y., Gutowsky, H. S., and Oldfield, E., 1979b, *Biochemistry* **18**:3268.

Kang, S. Y., Kinsey, R. A., Rajan, S., Gutowsky, H. S., Gabridge, M. G., and Oldfield, E., 1981, *J. Biol. Chem.* **256**:1155.

Ladbrooke, B. D., Williams, R. M., and Chapman, D., 1968, *Biochim. Biophys. Acta* **150**:333.

Larsson, K., 1977, *Chem. Phys. Lipids.* **20**:225.

Lentz, B. R., Barrow, D. A., and Hoechli, M., 1980, *Biochemistry* **19**:1943.

Levine, Y. K., Birdsall, N. J. M., Lee, A. G., and Metcalfe, J. C., 1972a, *Biochemistry* **11**:1416.

Levine, Y. K., Partington, P., Roberts, G. C. K., Birdsall, N. J. M., Lee, A. G., and Metcalfe, J. C., 1972b, *FEBS Lett.* **23**:203.

Lewis, B. A., DasGupta, S. K., and Griffin, R. G., 1984, *Biochemistry* **23**:1988.

Lewis, B. A., Harbison, G. S., Herzfeld, J., and Griffin, R. G., 1985, *Biochemistry* **24**:4671.

Mabrey, S., Mateo, P. L., and Sturtevant, J. M., 1978, *Biochemistry* **17**:2464.

McAlister, J., Yathindra, N., and Sundaralingam, M., 1973, *Biochemistry* **12**:1189.

McIntosh, T. J., 1980, *Biophys. J.* **29**:237.

Mehring, M., 1983, *Principles of High Resolution NMR in Solids*, 2nd edition, Springer-Verlag, Berlin.

Meier, P., Ohmes, E., Kothe, G., Blume, A., Weidner, J., and Eibl, H., 1983, *J. Phys. Chem.* **87**:4904.

Meier, P., Ohmes, E., and Kothe, G., 1986, *J. Chem. Phys.* **85**:3592.

Meulendijks, G. H. W. M., van Es, W., de Haan, J. W., and Buck, H. M., 1986, *Eur. J. Biochem.* **157**:421.

Müller, K., Meier, P., and Kothe, G., 1985, *Progr. Nucl. Magn. Reson. Spectrosc.* **17**:211
Oldfield, E., Gilmore, R., Glaser, M., Gutowsky, H. S., Hsung, J. C., Kang, S. Y., King, T. E., Meadows, M., and Rice, D., 1978, *Proc. Natl. Acad. Sci. U.S.A.* **75**:4657.
Pace, R. J., and Chan, S. I., 1982a, *J. Chem. Phys.* **76**:4228.
Pace, R. J., and Chan, S. I., 1982b, *J. Chem. Phys.* **76**:4241.
Pascher, I., and Sundell, S., 1986, *Biochim. Biophys. Acta* **855**:68.
Pauls, K. P., MacKay, A. L., Söderman, O., Bloom, M., Tanjea, A. K., and Hodges, R. S., 1985, *Eur. Biophys. J.* **12**:1.
Peterson, N. O., and Chan, S. I., 1977, *Biochemistry* **16**:2657.
Rice, D. M., Blume, A., Herzfeld, J., Wittebort, R. J., Huang, T. H., DasGupta, S. K., and Griffin, R. G., 1981, in: *Proceedings of the Second SUNYA Conversation in the Discipline Biomolecular Stereodynamics* (R. H. Sarma, ed.), Vol. II, pp. 255–270, Adenine Press, New York.
Rubinstein, J. L. R., Smith, B. A., and McConnell, H. M. 1979, *Proc. Natl. Acad. Sci. U.S.A.* **76**:15.
Ruocco, M. J., and Shipley, G. G., 1982, *Biochim. Biophys. Acta* **684**:59.
Ruocco, M. J., Siminovitch, D. J., and Griffin, R. G. 1985a, *Biochemistry* **24**:2406.
Ruocco, M. M., Makriyannis, A., Siminovitch, D. J., and Griffin, R. G., 1985b, *Biochemistry* **24**:4844.
Rüppel, D., and Sackmann, E., 1983, *J. Phys. (Paris)* **44**:1025.
Sackmann, E., 1978, *Ber. Bunsenges, Phys. Chem.* **82**:891.
Sackmann, E., and Träuble, H., 1972, *J. Am. Chem. Soc.* **94**:4492.
Seelig, A., and Seelig, J., 1975, *Biochim. Biophys. Acta* **406**:1.
Seelig, J., 1977, *Q. Rev. Biophys.* **10**:353.
Seelig, J., and Gally, H. U., 1976, *Biochemistry* **15**:5199.
Seelig, J., and Niederberger, W., 1974, *Biochemistry* **13**:1585.
Seelig, J., and Seelig, A., *Biochem. Biophys. Res. Commun.* **57**:406.
Seelig, J., and Seelig, A., 1980, *Q. Rev. Biophys.* **13**:19.
Seelig, J., Gally, H. U., and Wohlgemuth, R., 1977, *Biochim. Biophys. Acta* **467**:109.
Shaw, D., 1984, *Fourier Transform NMR Spectroscopy*, 2nd Edition, Elsevier, Amsterdam.
Slichter, C. P., 1978, *Principles of Magnetic Resonance*, Springer-Verlag, Berlin.
Smith, I. C. P., 1979, *Can. J. Biochem.* **57**:1.
Smith, I. C. P., and Jarrell, H. C., 1983, *Acc. Chem. Res.* **16**:266.
Smith, R. L., and Oldfield, E., 1984, *Science* **225**:280.
Solomon, I., 1958, *Phys. Rev.* **110**:61.
Spiess, H. W., 1978, *NMR: Basic Principles and Progress*, Vol. 15, p. 55, Springer-Verlag, Berlin.
Spiess, H. W., 1985, *Adv. Polym. Sci.* **66**:21.
Spiess, H. W., and Sillescu, H., 1981, *J. Magn. Reson.* **42**:381.
Stamatoff, J., Feuer, B., Guggenheim, H. J., Tellez, G., and Yamane, T., 1982, *Biophys. J.* **38**:217.
Strenk, L. M., Westerman, P. W., and Doane, J. W., 1985, *Biophys. J.* **48**:765.
Sugar, I. P., and Monticelli, G., 1983, *Biophys. Chem.* **18**:281.
Tardieu, A., Luzzatia, V., and Reman, F. C., 1973, *J. Biol. Chem.* **75**:711.
Wittebort, R. J., Schmidt, C. F., and Griffin, R. G., 1981, *Biochemistry* **20**:4223.
Wittebort, R. J., Blume, A., Huang, T. H., DasGupta, S. K., and Griffin, R. G., 1982, *Biochemistry* **21**:3487.

Chapter 3

Applications of Calorimetry
to Lipid Model Membranes

Alfred Blume

1. INTRODUCTION

Thermal techniques have found wide application in the study of properties of biological and lipid model membranes. Particularly differential scanning calorimetry (DSC) has become a standard technique for studying the
thermally induced transition of biological membranes and bilayers of model
lipids from an ordered, crystallinelike state at low temperature (gel phase)
to a more disordered liquid-crystalline state at higher temperature (Chapman *et al.*, 1967; Chapman, 1968; Ladbrooke and Chapman, 1969; Lee,
1977a, b; Barisas and Gill, 1978; Mabrey and Sturtevant, 1978; McElhaney,
1982; Bach, 1984; Donovan, 1984). This gel to liquid-crystalline phase
transition is a highly cooperative transition. The endothermic effect observed
on heating the system through the phase transition is caused by changes in
internal energy of the system as the number of *gauche* conformations in
the fatty acyl chains increases, and by concomitant changes in van der
Waals interactions between the chains and polar interactions at the lipid
bilayer–water interface. The "melting" of the fatty acyl chains is accompanied by an increase in total volume of the bilayer by ca. 2%–4% (Sheetz
and Chan, 1972; Blazyk *et al.*, 1975; Melchior and Morowitz, 1972), the

ALFRED BLUME • Institut für physikalische Chemie der Universität Freiburg, D-7800
Freiburg, West Germany.

lateral expansion in the plane of the bilayer being much larger (ca. 20%-
25%). This increase in molecular area at the bilayer-water interface is thus
compensated to a large extent by a thinning of the membrane in the
liquid-crystalline state.

   Thermodynamic data like transition temperatures, transition enthalpies, and the behavior of lipid mixtures have been determined for a large
number of synthetic and natural lipids using the DSC method. In addition
the influence of protein incorporation on the thermotropic behavior was
studied for a variety of reconstituted systems (McElhaney, 1982; Bach,
1984).

   A second calorimetric technique, reaction calorimetry, has been used
only in a limited number of cases. With this technique heats of ion binding
to charged phospholipids can be determined (Rialdi and Raffanti, 1984).
As the transition temperature of negatively charged lipids is dependent on
the head group charge, the transition from the liquid-crystalline state can
be induced isothermally by ion binding (Träuble and Eibl, 1974), so that
in addition to the binding enthalpy the transition enthalpy of the isothermally induced transition can be measured. These experiments can then be
performed as a function of temperature and from the temperature dependence of the reaction enthalpy the change in heat capacity of the reaction
can be determined (Rialdi and Raffanti, 1986).

   In certain cases, i.e. with amphiphilic peptides, reaction calorimetry
can also be employed to study the incorporation of small proteins into
membranes (Massey et al., 1981; Epand and Sturtevant, 1981). The heats
of incorporation observed can then be compared with the behavior of the
same lipid-protein systems in the DSC experiment. This gives information
on the thermodynamic state of the lipids in the annulus surrounding the
protein.

   This chapter gives a brief outline of the principles of these two
calorimetric methods and presents some recent examples for the application
of DSC and reaction calorimetry to model membrane systems. No comprehensive review of thermodynamic data of model membranes is intended.
The reader is referred to several reviews that have appeared in the literature
(Mabrey and Sturtevant, 1978; McElhaney, 1982; Bach, 1984; Donovan,
1984; Eibl, 1984).

2. DIFFERENTIAL SCANNING CALORIMETRY (DSC)

2.1. Principle

   The principle of differential scanning calorimetry is very simple. In a
DSC instrument two identical cells, one containing the sample solution and

the other an inert reference solution (for instance a buffer), are heated at a programmed rate. A control circuit using a thermopile to detect temperature differences between the two cells feeds differential heating power to either one of the cells in case there is a temperature difference between them. Initially the temperature of both cells increases linearly with time and the temperature difference between them is zero. When a thermally induced transition takes place in the sample solution the control circuit senses the resulting temperature difference between the two cells and additional power is fed to the sample cell to maintain a zero temperature difference. This differential heating power is recorded as a function of time or temperature on an x-y recorder and provides the basis for the determination of transition temperatures and transition enthalpies.

The total electric power N fed to either one of the cells results in an increase in temperature dT in a time interval dt when the heat capacity of the cell is c:

$$c = N\left(\frac{dt}{dT}\right) \tag{1}$$

As mentioned above only the differential power or differential heat capacity is of interest:

$$c_{\text{diff}} = c_{\text{sample}} - c_{\text{ref}} = (N_{\text{sample}} - N_{\text{ref}})\left(\frac{dt}{dT}\right) = N_{\text{diff}}\left(\frac{dt}{dT}\right) \tag{2}$$

In an ideal instrument c_{diff} is zero when no transition is taking place. When a thermal event like the endothermic lipid phase transition occurs, differential power N_{diff} is fed to the sample cell and c_{diff} is nonzero till the transition is over.

If we assume a simple cooperative two-state transition between the species A and B:

$$A \rightleftharpoons B \tag{3}$$

then we can write for the equilibrium constant K:

$$K = [B]/[A] = \theta/(1 - \theta) \tag{4}$$

with θ being the degree of transition, which is related to K by

$$\theta = K/(1 + K) \tag{5}$$

The temperature dependence of K is described by the van't Hoff equation:

$$K(T) = K(T_m) \exp[-\Delta H_{\text{v.H.}}/R \cdot (1/T - 1/T_m)] \tag{6}$$

From these equations the temperature dependence of c_{diff} can be calculated:

$$c_{diff} = \Delta H_{cal}\left(\frac{d\theta}{dT}\right) = \Delta H_{cal}\left(\left\{1 + \exp\left[\frac{-\Delta H_{v.H.}}{R}\left(\frac{1}{T} - \frac{1}{T_m}\right)\right]\right\}^{-2}\right.$$

$$\left. \times \exp\left[\frac{-\Delta H_{v.H.}}{R}\left(\frac{1}{T} - \frac{1}{T_m}\right)\right]\frac{\Delta H_{v.H.}}{RT^2}\right) \tag{7}$$

where ΔH_{cal} is the molar transition enthalpy determined calorimetrically and $\Delta H_{v.H.}$ is the van't Hoff transition enthalpy calculated from the temperature dependence of K, R is the gas constant, and T_m is the transition temperature (Engel and Schwarz, 1970; Träuble, 1971). Figure 1 shows a calculated DSC curve for a transition at 314 K with a ΔH_{cal} value of 8.7 kcal mol^{-1} and a van't Hoff enthalpy of 1500 kcal mol^{-1}. In the calorimetric experiment the recorded curves look very similar to the one shown. In these experiments the heats of transition ΔH_{cal} and $\Delta H_{v.H.}$ are unknowns. The first step is to determine the total heat of transition Δh_t by measuring the area under the peak between the temperatures where c_{diff} departs from the base line:

$$\Delta h_t = \int_{t_{onset}}^{t_{end}} N_{diff}\, dt = \int_{T_{onset}}^{T_{end}} c_{diff}\, dT \tag{8}$$

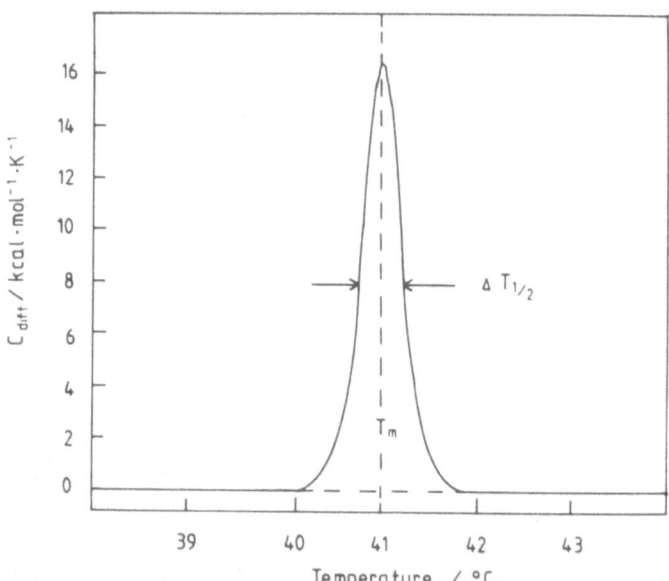

Figure 1. Theoretical calorimetric curve, calculated for a two-state transition according to equation (3) with the following parameters: $T_m = 314$ K, $\Delta H_{cal} = 8.7$ kcal mol^{-1}, $\Delta H_{v.H.} = 1500$ kcal mol^{-1}.

From Δh_t the molar transition enthalpy ΔH_{cal} can be calculated from the known mass of material and the molecular weight. For cooperative transitions like the lipid phase transition where intermolecular cooperativity is involved the calorimetrically determined transition enthalpy ΔH_{cal} is much lower than the van't Hoff enthalpy $\Delta H_{v.H.}$. The ratio $\Delta H_{v.H.}/\Delta H_{cal}$ is thus larger than unity. This ratio is usually called the cooperative unit size c.u. (Engel and Schwarz, 1970; Träuble, 1971) and is a measure for the degree of cooperativity between the lipid molecules. For the case of a true first-order transition this ratio approaches infinity and for the case of a completely noncooperative equilibrium this ratio is unity. To determine c.u. $\Delta H_{v.H.}$ has to be known. This can be calculated easily from equation (7) at $T = T_m$ where $\theta = 1/2$ (Engel and Schwarz, 1970; Träuble, 1971):

$$(d\theta/dT)_{T_m} = \Delta H_{v.H.}/(4RT_m^2) \tag{9}$$

and

$$\Delta H_{v.H.} = 4RT_m^2(d\theta/dT) \tag{10}$$

Thus the slope of the transition curve evaluated by integration according to equation (8) at $T = T_m$ is used to calculate $\Delta H_{v.H.}$. So from one calorimetric experiment the transition enthalpies ΔH_{cal} and $\Delta H_{v.H.}$ and the cooperative unit size c.u. can be determined. $\Delta H_{v.H.}$ can also be estimated directly from the half height width $\Delta T_{1/2}$ of the c_{diff} curve (Mabrey and Sturtevant, 1978):

$$\Delta H_{v.H.} \approx 6.9 \, T_m^2/\Delta T_{1/2} \tag{11}$$

This expression is accurate to within 10% of the real value (Mabrey and Sturtevant, 1978).

2.2. Instrumentation

The number of commercially available differential scanning calorimeters suitable for studies of dilute aqueous solutions or suspensions is still very limited. One of the most widely used instruments is the Perkin-Elmer DSC family (Perkin-Elmer Corporation, Norwalk, Connecticut, USA). These instruments were originally built for investigations of solid samples but can also be used with volatile samples using different sample pans. The major drawback of this type of instrument is the relatively high concentration needed (5–10 wt%). As the sample volume is only 20–50 μl the total amount of sample required is low. For high-precision measurements the differential scanning calorimeters of the adiabatic type are to be preferred. These instruments require much lower concentrations (0.02–0.1 wt%) but have a larger fixed sample volume (ca. 1 ml). In addition very slow scan rates (down to 0.05 deg/min or even less) can be used. In these instruments the two cells are surrounded by one or two adiabatic shields, which are kept at the same temperature as the cells. Thus heat flow from or to the cells is prevented, resulting in higher sensitivity. Two instruments of this type are commercially available, the Microcal MC1 (Amherst, Massachusetts, USA) (Krishan and Brandts, 1978), and the so-called Privalov calorimeter DASM-1M with its newer version DASM-4M (Mashpriborintorg, Moscow, USSR) (Privalov *et al.*, 1975). A schematic diagram of the cell assembly of the DASM-1M is shown in Fig. 2. The use of fixed cells with good mechanical stability and a highly symmetric design gives an excellent base-line stability and reproducibility, which is essential for determining transition enthalpies of very broad peaks with sufficient accuracy. For this instrument the reproducibility of the base line after refilling the cells is ca. ± 25 μcal K^{-1} cm^{-3} for a scan rate of 1 K min^{-1}. This excellent performance makes it possible to determine specific heats of dissolved or dispersed compounds at relatively low concentrations (1–2 mg/ml) from the displacement of the base line of a scan with a sample relative to the base line when only buffer is in the sample cell (Privalov and Khechinashvili, 1974).

Converting the analog signals for temperature and differential heating power into digital form with an A/D converter, data acquisition on-line or off-line with subsequent computer analysis can easily be accomplished (Blume, 1983). Integration of equation (8) can then be performed digitally. Otherwise the area under the peak has to be determined by means of a planimeter or by paper cutting and weighing.

2.3. Applications of DSC to Lipid Model Membrane Studies

2.3.1. Determination of T_m and ΔH of Single Lipids

Differential scanning calorimetry has become the most widely used technique to study the phase behavior of lipids. Since the pioneering work

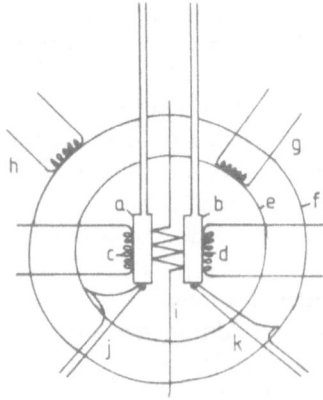

Figure 2. Schematic diagram of the cell assembly of the Privalov calorimeter (Privalov *et al.*, 1975): (a) sample cell; (b) reference cell; (c) and (d) heaters for sample and reference cell, respectively; (e) inner adiabatic shield; (f) outer adiabatic shield; (g) and (h) heaters for inner and outer adiabatic shield; (i) thermopile between sample and reference cell; (j) and (k) thermopiles between inner adiabatic shield and cells and outer adiabatic shield and cells, respectively.

of D. Chapman and co-workers in the late 1960s (Chapman *et al.*, 1967; Chapman, 1968; Ladbrooke and Chapman, 1969) a large number of studies have been conducted investigating the phase behavior of synthetic lipids in aqueous dispersion. Most experiments have been performed with phospholipids, as this is one of the most abundant lipid classes in biological membranes. Anhydrous phospholipids show thermotropic mesomorphism, i.e., several phase changes occur below the capillary melting point (Chapman, 1965). These compounds can thus exist in intermediate liquid-crystalline states below the isotropic melt. In addition they exhibit lyotropic mesomorphism, i.e., their phase state changes upon addition of water. As an example, Fig. 3 shows DSC traces of distearoyl-phosphatidylcholine (DSPC) with increasing amount of water (Chapman *et al.*, 1967). The transition temperature for the transition from the crystalline to the liquid-crystalline state decreases with water content up to a certain concentration.

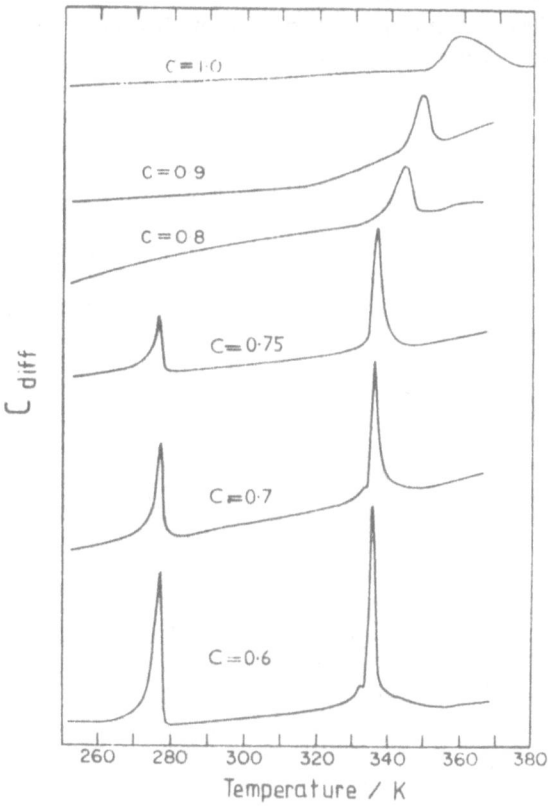

Figure 3. Calorimetric scans of DSPC with increasing amounts of water; c is the weight fraction of DSPC (after Chapman, 1968).

Adding excess water does not lead to any further changes, only the peak due to the melting of ice at 0°C increases. In the presence of excess water the lipids are arranged in bilayers separated to a certain distance by water layers, the excess water being free water not located in between the bilayers. An appreciable number of these water molecules in between the bilayers are "bound" to the head group. This "bound" water is different from free water in quite a few of its properties, for instance it does not freeze at 0°C. The total amount of "bound" water depends on the chemical structure of the lipid head group and on its net electric charge (Hauser, 1975; Cevc and Marsh, 1985). The amount of "bound" water also depends on the phase state of the lipids, being larger above the transition temperature when the molecules in the bilayer are in the liquid-crystalline state.

In this phase the lipid molecules can diffuse rapidly in the plane of the bilayer, and in addition fast rotational diffusion and *trans-gauche* isomerization in the fatty acyl chains is taking place. Cooling the system through the transition temperature results in a change of the state of the molecules. They become quasicrystalline with their chains extended in an almost all-*trans* conformation, and the molecules are now tightly packed

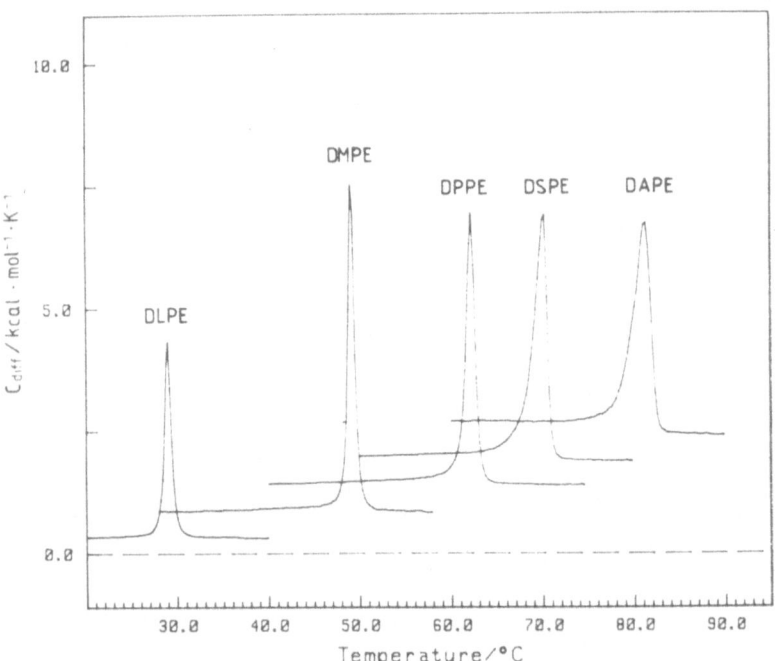

Figure 4.　Calorimetric scans of phosphatidylethanolamines (PEs) with different chain lengths ($c = 1$ mg/ml).

in a hexagonal or quasihexagonal lattice. All motions are slowed down to a large extent.

As all biological membranes are in general in excess water, we will now discuss only those parameters influencing the phase transition properties of phospholipids at high water content.

Figure 4 shows DSC scans of a series of phosphatidylethanolamines with saturated fatty acyl chains differing only in the length of the chains. An increase in chain length causes an increase in transition temperature T_m and transition enthalpy ΔH. From ΔH and T_m the transition entropy $\Delta S = \Delta H / T_m$ can be calculated as $\Delta G = 0$ at T_m. The ΔS values calculated that way for an additional CH_2 group are much lower than the entropy change observed for the melting of pure hydrocarbons or long-chain fatty acids, indicating that the bilayer still has a relatively high degree of order in the liquid-crystalline state. Thus this phase is drastically different compared to an isotropic solution. Table I summarizes the thermodynamic data derived for a variety of phospholipids in excess water.

Table I.   Transition Temperatures for Pretransition (T_{m_1}) and Main Transition (T_{m_2}), Transition Enthalpies for Pre-(ΔH_1) and Main Transition (ΔH_2), and Transition Entropies for Main Transition (ΔS_2) for Phospholipids with Identical Chains in Dilute Aqueous Dispersion[a]

Lipid	T_{m_1} (°C)	T_{m_2} (°C)	ΔH_1 (kcal mol^{-1})	ΔH_2 (kcal mol^{-1})	ΔS_2 (cal mol^{-1} K^{-1})
DMPC	15.3	24.0	1.3	6.5	21.9
DPPC	35.5	41.5	1.6	8.7	27.7
DSPC	51.0	54.3	1.8	10.9	33.3
DAPC	62.1	64.1	1.7	12.3	37.6
DBPC		72.5		14.9	43.1
DLPE		30.5		4.3	14.2
DMPE		49.9		6.6	20.4
DPPE		63.9		8.6	25.5
DSPE		72.5		10.5	30.4
DAPE		81.1		12.2	34.5
DLPA (pH 6)		33.5		3.4	11.1
DMPA (pH 6)		52.2		5.7	17.5
DPPA (pH 6)		65.0		7.9	23.4
DMPA (pH 12)		22.4		4.1	13.9
DPPA (pH 12)		43.1		5.7	17.8
DHPC	33.0	43.5	1.4	8.5	26.7
DHPE		68.5		7.6	22.3
DHPA (pH 6)		73.3		7.1	20.5
DHPA (pH 12)		53.8		5.8	17.8

[a] Blume (1983 and unpublished).

PEs are simple systems in the sense that under low ionic strength conditions they show only one thermotropic event, i.e., a change from the lamellar L_β phase at low temperature to the liquid-crystalline L_α phase (McIntosh, 1980), though they tend to dehydrate upon standing at low temperature, converting to a crystalline phase (Seddon *et al.*, 1983). Other phospholipids like phosphatidylcholines (PC), which have been studied in much detail, have an additional transition between two types of ordered phases, the so-called pretransition, which has a much lower transition enthalpy. At the pretransition the lamellar $L_{\beta'}$ phase, where the chains are tilted with respect to the bilayer normal, transforms to the P_β or $P_{\beta'}$ phase, where the bilayer surface displays a periodic ripple structure (Chapman *et al.*, 1967). Recent NMR experiments have revealed that the P_β phase is actually microscopically heterogeneous as two spectroscopically different components can be observed in the temperature range between the pre- and main transition (Wittebort, 1981). PCs even show a third phase transition at low temperature when the hydrated $L_{\beta'}$ phase is dehydrated and a crystalline-like lamellar L_c phase is formed. This $L_{\beta'} \rightarrow L_c$ transition shows a large hysteresis and is kinetically hindered. For DPPC the bilayers have to be kept at 0°C for several days before the L_c phase is formed. Raising the temperature again will induce the $L_c \rightarrow L_{\beta'}$ transition at 17°C with a transition enthalpy much larger than that of the pretransition at 35°C (Chen *et al.*, 1980).

The thermodynamic parameters ΔH, ΔS, and T_m for the main transition of PCs are also chain length dependent, similarly to the PEs. However, the transition temperatures of PCs are lower and the transition enthalpies are somewhat higher when compounds with identical chain lengths are compared (see Table I). This is an indication for the existence of stronger head group interaction in the case of PEs. Hydrogen bonds between the positively charged $-NH_3^+$ groups and neighboring negatively charged phosphate groups may be responsible for this effect (Eibl and Woolley, 1979).

As can be seen from Table I, the increments in ΔH per methylene group are ca. 0.5 cal mol^{-1} for PEs and 0.54 cal mol^{-1} for PCs. The increments for ΔS per methylene are also slightly different for these two lipid classes, namely, 1.31 and 1.37 cal K^{-1} mol^{-1} for PEs and PCs, respectively. This may again be an indication of a tighter packing of the PE molecules in the liquid-crystalline phase caused by the attractive head group interactions via hydrogen bonds, which are not possible for PCs. Other phospholipids like phosphatidylglycerols (PG) and phosphatidic acids (PA) show a similar thermotropic behavior. In particular PGs at pH values above 7 behave very similarly to PCs, and the thermotropic behavior of PAs in their singly charged form at pH 7 is similar to that of the corresponding PEs.

Biological phospholipids generally have two different fatty acids esterified to the two hydroxyl groups of the glycerol backbone, unsaturated fatty acids being preferentially linked to the 2-position of the glycerol. The transition characteristics of the two isomers of a phospholipid with mixed acyl chains can be quite different. In the case of saturated PCs, the isomer having the longer fatty acid in the 2-position will have the higher transition temperature and transition enthalpy for the main transition compared to the isomer with reversed orientation of the two chains. For instance, PMPC has a T_m of 27°C and a ΔH value of 6.5 kcal mol^{-1}. However MPPC has a T_m of 37°C and ΔH value of 7.3 kcal mol^{-1} (Stümpel et al., 1981; 1983). Elongation of the 2-chain fatty acid with a constant chain length of the other chain leads to an increase in ΔH and ΔS with an increment per CH_2 group that is not constant but decreases with increasing chain length from ca. 1.5 to 0.2 kcal mol^{-1} for ΔH and 4.6 to 0.6 cal K^{-1} mol^{-1} for ΔS. Table II shows data for a series of PCs with a palmitoyl chain in the 1-position and different fatty acids in the 2-position (Stümpel et al., 1981; 1983). When the fatty acid in the 2-position is kept constant and the other fatty acid is varied T_m increases much less with chain length and ΔH stays almost invariant (see Table II) (Stümpel et al., 1981; 1983; Mason et al., 1981). A similar behavior is found for the corresponding PEs, only that in this case the transition temperatures are higher as expected. Table II includes some data for this phospholipid class (Hübner and Blume, unpublished).

The effects of different chains lengths in the two positions can be explained by the particular conformation of the glycerol backbone of a phospholipid molecule. The conformation is such that the chain in the 2-position is bent at the first two methylene groups of the chains as it first starts almost parallel to the plane of the bilayer before becoming perpendicular to the surface. The result of this conformation is that the effective

Table II. Transition Temperatures (T_{m_2}), Transition Enthalpies (ΔH_2), and Transition Entropies (ΔS_2) for the Main Transition for Phospholipids with Mixed Chain Lengths in Dilute Aqueous Dispersion[a]

Lipid	T_{m_2} (°C)	ΔH_2 (kcal mol^{-1})	ΔS_2 (cal mol^{-1} K^{-1})
PLPC	9.6	3.4	12.0
PMPC	28.0	6.4	21.3
DPPC	41.5	8.7	27.7
PSPC	48.8	10.3	31.3
PAPC	51.4	11.6	35.7
PBPC	52.1	12.0	35.8
DPPE	63.9	8.6	25.5
PSPE	69.4	9.6	28.0
PAPE	72.4	11.9	34.4

[a] Hübner and Blume (unpublished).

length of the 2-chain is reduced so that the terminal methyl groups are not in register (Hitchcock *et al.*, 1974).

2.3.2. Determination of Apparent Molar Beat Capacities of Phospholipids

Differential scanning calorimetry can not only be used to determine the transition properties of phospholipid bilayers, i.e., the changes in enthalpy at the phase transition, but also to measure the heat capacities of the dispersed lipid itself (Blume, 1983; Wilkinson and Nagle, 1982). A prerequisite for this type of measurement is a DSC instrument with a highly reproducible and stable base line (see Section 1.3.1). The apparent molar heat capacity $^{\phi}C_p$ of a dispersed lipid can be calculated according to the procedure described by Privalov and Khechinashvili (1974) using the following equation:

$$^{\phi}C_p = [c_{p_w}(V_L/V_w) - \Delta/m_L]M_L \tag{12}$$

with M_L the molecular weight of the lipid, m_L the mass of lipid in the sample cell, V_L and V_w the specific volume of lipid and water, respectively, c_{p_w} the specific heat of water, and Δ the displacement of the base line (in cal K^{-1}) in a DSC run with the sample in the cell compared to the base line recorded with buffer in both cells.

Figure 5 shows a schematic diagram of the base line shift observed in the DSC experiment when the sample cell is loaded with a lipid dispersion.

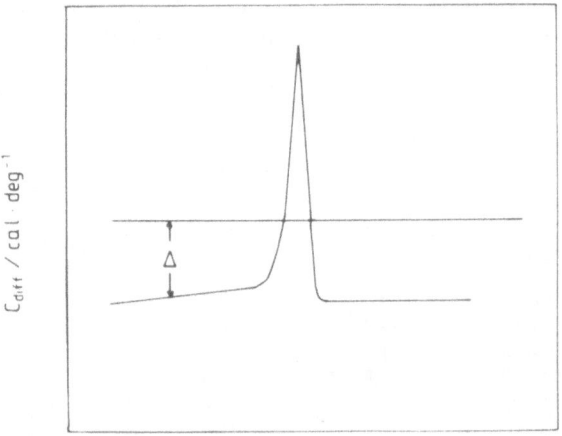

Temperature / °C

Figure 5. Schematic drawing of the calorimetric scans used for the calculation of apparent molar heat capacities $^{\phi}C_p$. Upper curve: base line (water–water); lower curve: experimental curve (lipid dispersion–water); Δ: displacement in cal deg^{-1}.

The values for c_{p_w} and V_W can be taken from standard tables. V_L has to be measured separately by densitometry. The characteristic range of the specific volumes of PCs and PEs in the gel and liquid-crystalline phases is between 0.9 and 1.02 ml g^{-1}, higher values applying to the liquid-crystalline phase as the volume change at the transition is between 2% and 4% (Sheetz and Chan, 1972; Blazyk *et al.*, 1975; Melchior and Morowitz, 1972; Nagle and Wilkinson, 1978; Wilkinson and Nagle, 1981).

Figure 6 shows some representative scans of PCs and Table III summarizes some data for different phospholipids (Blume, 1983). The transition peaks in Fig. 6 are off-scale to give a better resolution for $^\phi C_p$. As can be seen, $^\phi C_p$ increases with chain length. The absolute values of $^\phi C_p$ contain important information on lipid–water interactions. Generally the hydration of polar groups will give negative contributions to the total heat capacity arising from the orientation of the water molecules surrounding the polar groups. However, "hydrophobic hydration" of apolar parts like methylene groups gives positive contributions to the total heat capacity because of entropic effects. The $^\phi C_p$ values of phospholipids are larger than approximated for a bilayer system where only the head groups including the glycerol backbone are in contact with water. The approximation can be performed in two different ways. Firstly, the apparent molar heat capacity of a micellar system like 1-palmitoyl-lyso-phosphatidylcholine (LPC) can be measured (see Fig. 6). In the case of micelles it is known from other types of experiments that "hydrophobic hydration" plays a role. It is estimated that four to five methylene groups of a surfactant molecule are in contact with

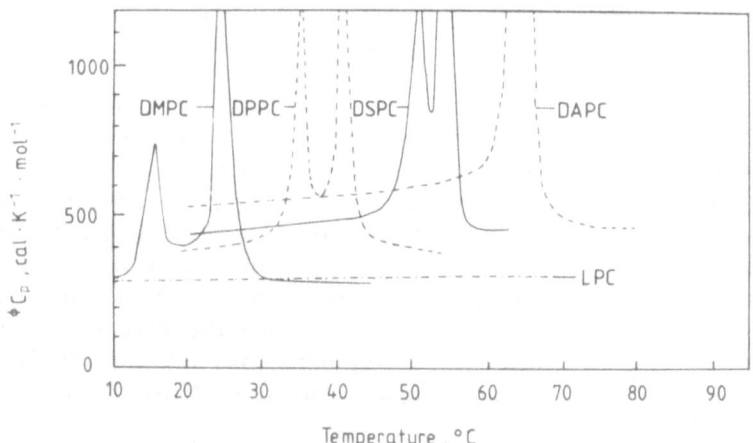

Figure 6. DSC scans for phosphatidylcholines (PCs) with different chain lengths. The transition peaks are off-scale in this presentation to show the apparent molar heat capacities $^\phi C_p$. Included is a curve for the micelle forming lipid 1-palmitoyl-lyso-phosphatidylcholine (LPC) (Blume, 1983).

Table III. Apparent Molar Heat Capacities $^{\phi}C_p$ in cal mol^{-1} K^{-1} of Phospholipids at Temperatures $T - T_m = \Delta T = -20°C$, i.e. in the Gel Phase and at $\Delta T = +10°C$ in the Liquid-Crystalline Phase (Blume, 1983)

Lipid	$\Delta T = -20°C$	$\Delta T = +10°C$
DMPC	270	305
DPPC	380	410
DSPC	470	450
DAPC	580	480
DLPE	260	305
DMPE	350	355
DPPE	430	400
DSPE	480	400
DAPE	560	445
DMPA (pH 6)	380	250
DPPA (pH 6)	426	290
DMPA (pH 12)	295	380
DPPA (pH 12)	412	420

water (Kresheck, 1975; Menger, 1979). To approximate the heat capacity of a double chain phospholipid the value for the heat capacity of a liquid or solid alkane chain is added. For the other type of approximation group parameter values for the heat capacity of hydrated groups as reported in the literature are added in an appropriate way to get an estimate for $^{\phi}C_p$ for a phospholipid (Nichols et al., 1976; Roux et al., 1978). In both cases the approximated values for the heat capacity of phospholipids in the liquid-crystalline phase are ca. 80–100 cal K^{-1} mol^{-1} lower than the values determined experimentally. This is an indication for contributions from "hydrophobic hydration," making it likely that water molecules not only surround the head groups but can also come into contact with methylene groups of the fatty acyl chains. Thermodynamic measurements, however, can give no information on the location of these water molecules, and the question is whether they are "dissolved" in the bilayer or whether water can penetrate deeper into the hydrocarbon region reaching in from both bilayer surfaces. In any case, the experimentally observed high heat capacities of lipids indicating "hydrophobic hydration" could give some explanation for the unusually high water permeability of lipid bilayers (Blume, 1983).

2.3.3. Determination of Phase Diagrams of Binary Lipid Mixtures

DSC is the standard and most sensitive technique to determine the mixing behavior of different lipids. In biological membranes the lipid composition is very complex. Lipids with different head groups and fatty acyl chains are mixed together. Experiments determining the miscibility of individual components of this complex system are therefore of great interest. Deviations from ideal mixing behavior for two different species or complete immiscibility can have important implications for a biological membrane, namely, the possibility of lateral inhomogeneities. The phase behavior of binary mixtures of two different phospholipids can be treated in much the same way as mixtures in a three-dimensional system (Lee, 1977b). Experimentally the determination of a phase diagram (the T-x diagram) from the DSC scans is the first step. Figure 7 shows DSC scans of DMPC-DMPE mixtures at various mixing ratios. These two lipids have the same chain length but different head groups, DMPE having the higher transition temperature. From these scans the temperature–composition diagram is constructed using the temperatures of the onset and end of "melting" and plotting them versus the mole fraction x of DMPE. These temperatures are operationally defined as the points in the DSC scans where the c_{diff} values depart from the linear part of the base line below and above the phase transition. As can be seen from the scans in Fig. 7, these points are in some cases not easy to define because of the great width of the transition due to phase separation and additional broadening caused by reduced

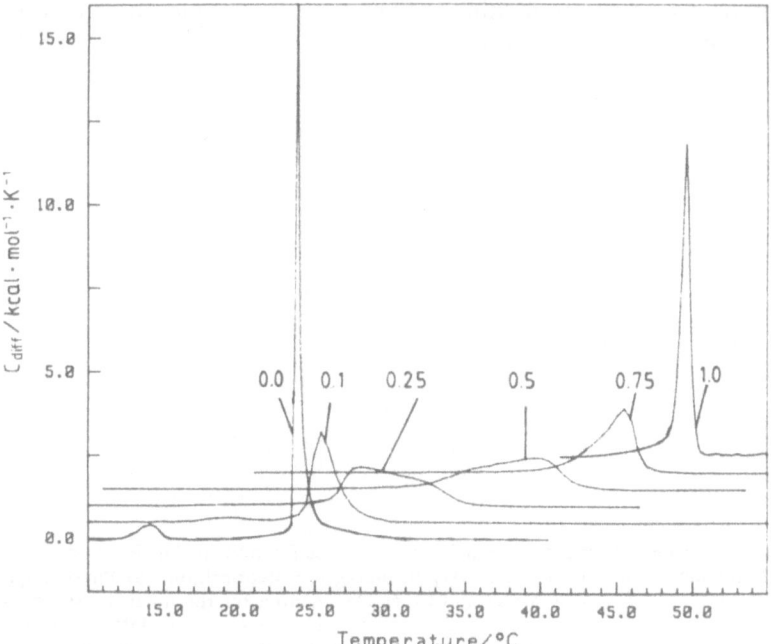

Figure 7.   DSC scans for DMPC–DMPE mixtures. Numbers designate mole fraction of DMPE.

cooperativity. The latter effect can be calculated and then removed giving a corrected phase diagram [see Eq. (25)].

Figure 8a shows the transition enthalpy as a function of composition for the DMPC–DMPE system. The uncorrected phase diagram in Fig. 8b is indicated by the dashed lines going through the experimental points. This binary mixture displays a typical phase diagram indicating complete miscibility in the gel as well as the liquid-crystalline phase (Chapman *et al.*, 1974). However, the different gel phase structures of the two components will give additional complications for the construction of the phase diagram in the low-temperature region. Pure DMPE has a L_β phase with the acyl chains perpendicular to the bilayer surface, while DMPC has an $L_{\beta'}$ or $P_{\beta'}$ phase. Thus the phase diagram at low temperature is actually more complicated and will show additional two-phase regions between $P_{\beta'}$ and $L_{\beta'}$ phases. For the following treatment of solid–liquid equilibria with nonideal mixing, this additional effect will be neglected.

Because of the difference in the nature of the head groups of the two lipids, ideal mixing behavior cannot be expected. These deviations from ideal behavior are due to differences in interaction energies between the two species. For the case of ideal behavior in both phases the composition of both phases in equilibrium, i.e., the points on the *liquidus* and *solidus* lines, can easily be calculated from the chemical potentials of the two species when their melting points and latent heats of melting are known. When component (1) has the melting temperature T_1 and melting enthalpy ΔH_1 and component (2) T_2 and ΔH_2, respectively, then it follows (see standard

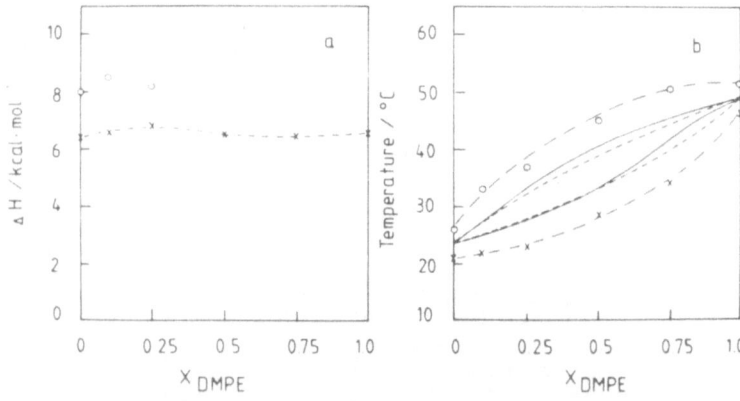

Figure 8. (a) Plot of transition enthalpy versus mole fraction for DMPC–DMPE mixtures; ×, ΔH for main transition only; ○, ΔH including the pretransition. (b) Phase diagram for DMPC–DMPE mixtures; ×, onset and ○, end of melting for the main transition as taken from the experimental curves in Fig. 7. Dashed line: phase diagram calculated for ideal mixing; solid line: phase diagram calculated for nonideal mixing with parameters as listed in Table IV.

text books of physical chemistry)

$$\ln\left(\frac{x_1^{liq}}{x_1^{solid}}\right) = \frac{\Delta H_1}{R}\left(\frac{T - T_1}{TT_1}\right) \tag{13}$$

$$\ln\left(\frac{x_2^{liq}}{x_2^{solid}}\right) = \ln\left(\frac{1 - x_1^{liq}}{1 - x_1^{solid}}\right) = \frac{\Delta H_2}{R}\left(\frac{T - T_2}{TT_2}\right) \tag{14}$$

These two equations can be solved for x_1^{liq} or x_1^{solid}, respectively, giving the points on the *liquidus* and *solidus* curves of the phase diagram. The phase diagram obtained for ideal mixing is indicated by the small dashed lines in Fig. 8b.

In the case of nonideal mixing in both phases the mole fractions x_i have to be replaced by the activities $a_i = x_i f_i$ with f_i being the activity coefficients. It thus follows from equations (13) and (14):

$$\ln\left(\frac{a_1^{liq}}{a_1^{solid}}\right) = \ln\left(\frac{x_1^{liq} f_1^{liq}}{x_1^{solid} f_1^{solid}}\right) = \frac{\Delta H_1}{R}\left(\frac{T - T_1}{TT_1}\right) \tag{15}$$

$$\ln\left(\frac{a_2^{liq}}{a_2^{solid}}\right) = \ln\left(\frac{x_2^{liq} f_2^{liq}}{x_2^{solid} f_2^{solid}}\right)$$

$$= \ln\left(\frac{(1 - x_1^{liq}) f_2^{liq}}{(1 - x_1^{solid}) f_2^{solid}}\right) = \frac{\Delta H_2}{R}\left(\frac{T - T_2}{TT_2}\right) \tag{16}$$

The activity coefficients f_i can be obtained from a series expansion of the molar excess free energy of mixing $\overline{\Delta G}^E$ according to Guggenheim (1937):

$$\overline{\Delta G}^E = x_2(1 - x_2)[A + B(2x_2 - 1) + C(2x_2 - 1)^2 + \cdots] \tag{17}$$

One obtains for the logarithm of the activity coefficients:

$$\ln f_1 = x_2^2 / RT[A + B(4x_2 - 3) + C(2x_2 - 1)(6x_2 - 5) + \cdots] \tag{18}$$

$$\ln f_2 = (1 - x_2)^2 / RT[A + B(4x_2 - 3) + C(2x_2 - 1)(6x_2 - 5) + \cdots] \tag{19}$$

For the simplest case of the so-called symmetrical mixture the coefficients $B, C, D, \ldots$ in equations (18) and (19) are zero, and only A is nonzero. This reduces equations (17), (18), and (19) to

$$\overline{\Delta G}^E = x_2(1 - x_2)A \tag{20}$$

$$\ln f_1 = x_2^2 A / RT \tag{21}$$

$$\ln f_2 = (1 - x_2)^2 A / RT \tag{22}$$

The nonideality of the system is thus contained in the coefficient A. As $\overline{\Delta G}^E = \overline{\Delta H}^E - T\overline{\Delta S}^E$ one can discuss the following two theoretical cases:

If $\overline{\Delta S}^E = 0$ then $\overline{\Delta G}^E = \overline{\Delta H}^E$. This is the case of the so-called regular solution (Hildebrand, 1929). It essentially means that A is temperature independent, because one can show that

$$\overline{\Delta S}^E = -(dA/dT)x_2(1 - x_2) \qquad (23)$$

The other extreme would be the case of the so-called athermal solution, where $\overline{\Delta H}^E = 0$ and thus $\overline{\Delta G}^E = \overline{\Delta S}^E$ (Guggenheim, 1944). In reality, of course, neither of these two cases exists. Regular solution theory predicts a random distribution of the two components in the mixture despite differences in interaction energies between the two species, whereas the athermal model puts all the nonideality in a nonrandom distribution, which is actually caused by a nonzero $\overline{\Delta H}^E$.

Independent of how the nonideality parameter is interpreted, for the determination of A the phase diagram has to be simulated with A as a variable parameter using equations (15) and (16), inserting the expressions for the activity coefficient from equations (21) and (22), and then compared to the experimental phase diagram. Because equations (15) and (16) are of transcendental nature, an iterative procedure has to be used (Lee, 1978).

As mentioned above an additional complication arises from the broadening of the DSC curves because of limited cooperativity, which makes the determination of the points on the *liquidus* and *solidus* curves very ambiguous. In our opinion an improved simulation procedure is to directly simulate the experimental DSC curves using a phase diagram calculated with a certain value for the nonideality parameter A. The DSC scans can be calculated from the phase diagrams according to the lever rule using equations first proposed by Mabrey and Sturtevant (1976):

$$c_p(T) = \frac{x_1^{liq}\Delta H_1 + x_2^{liq}\Delta H_2}{x_2^{solid} - x_2^{liq}} f^{solid} \frac{dx_2^{solid}}{dT}$$

$$+ \frac{x_1^{solid}\Delta H_1 + x_2^{solid}\Delta H_2}{x_2^{solid} - x_2^{liq}} f^{liquid} \frac{dx_2^{liq}}{dT} \qquad (24)$$

with $f^{solid} = (x_2 - x_2^{liq})/(x_2^{solid} - x_2^{liq})$ and $f^{liq} = 1 - f^{solid}$ being the fractions of lipid in the solid and liquid-crystalline state, respectively. However, this equation only applies to a system with a first-order transition with infinite cooperativity. The limited cooperativity of the real system can be included by convoluting the $c_p(T)$ function of equation (24) with a broadening function describing the width of the transition. This is essentially the temperature-dependent function $(d\theta/dT)$ described in equation (7), i.e., $c_{diff}/\Delta H_{cal}$. $\Delta H_{v.H.}$ is then an adjustable parameter determining the additional broadening. The final output function $c_{p_0}(T)$ is then obtained by the convolution

$$c_{p_0}(T) = \left(\frac{d\theta}{dT}\right)(T) \otimes c_p(T)$$

$$= \int_{-\infty}^{+\infty} \left(\frac{d\theta}{dT}\right)(t) c_p(T - t) \, dt \qquad (25)$$

The simulated c_{p_0} curves can now be compared with the experimental DSC scans, giving a more direct measure of the quality of the simulation, i.e., the correctness of the value of A used for the calculation of the phase diagram. The quality criteria using this procedure are not only the temperatures on the *liquidus* and *solidus* curves but in addition the similarities of the shapes of the DSC peaks calculated from the phase diagrams (Blume, unpublished).

Figure 9 shows simulated DSC scans for the DMPC–DMPE system in comparison with the experimental curves taken from Fig. 7. The phase diagram calculated from the simulation parameters is shown in Fig. 8b with

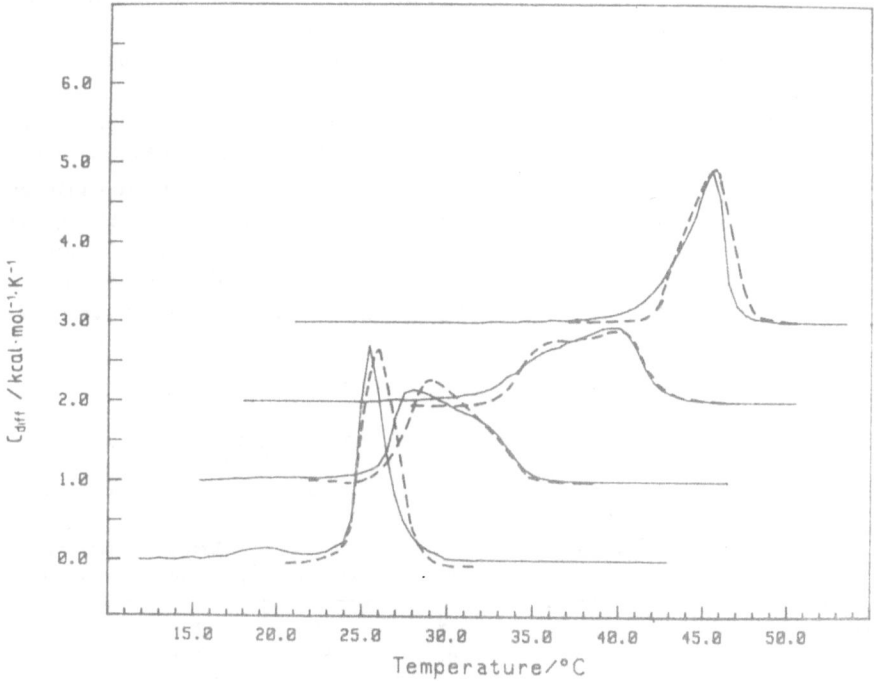

Figure 9. Solid lines: DSC curves for DMPC–DMPE mixtures (same as in Fig. 7); dashed lines: calculated calorimetric curves using the phase diagram for nonideal mixing shown in Fig. 8.

the solid lines. For this simulation it turned out to be necessary to use a temperature-dependent parameter A of the form $A = a - bT$ for both liquid and solid phases, a and b being characteristic constants. These parameters and the values derived for $\overline{\Delta G}^E$, $\overline{\Delta H}^E$, and $\overline{\Delta S}^E$ at a mole fraction of $x_2 = 0.5$ are shown in Table IV (Blume, unpublished). With these parameters the experimental DSC curves could be fitted quite well, as can be seen from Fig. 9. The high $\overline{\Delta H}^E$ value in the gel state could arise from the breaking of hydrogen bonds between the PE head groups when they are mixed with the PC head groups (Sugar and Monticelli, 1983). In the liquid-crystalline phase, this effect is expected to be smaller as the hydrogen bonds between the PE head groups are already weakened owing to the lateral expansion of the bilayer (Eibl and Wooley, 1979). Thus $\overline{\Delta H}^E$ is expected to be lower for the liquid-crystalline phase.

This example of a binary mixture of two phospholipids shows how in principle information on nonideal mixing behavior can be obtained from a simulation of the phase diagram and the DSC curves. As a consequence of nonideal mixing the formation of clusters of like molecules has to be expected. Estimations of this lateral species separation have been carried out by van Dreele (1978). Computer simulations using Monte Carlo methods by Freire and Snyder (1980) have shown that clusters formed of like molecules have cluster sizes of 25 to 40 molecules, getting larger with increasing nonideality of the mixture.

In the case of phospholipid mixtures with one of the components being a negatively charged species like PA, PG, or PS, the mixing behavior can be modified by binding of positive ions to the negatively charged groups. An example of this case is the DMPC–DMPA mixture shown in Fig. 10 (Blume, unpublished). The phase diagram for this binary system is similar to the one obtained for DMPC–DMPE. However, binding of Ca^{2+} to the negatively charged DMPA molecules can modify the mixing behavior. For pure dilute DMPA dispersions $1:2$ binding of Ca^{2+} increases the transition

Table IV. Simulation Parameters for DSC Curves of DMPC–DMPE Mixtures as Described in the Text[a]

	a (cal mol^{-1})	b (cal mol^{-1} K^{-1})	A_{298} (cal mol^{-1})	$\overline{\Delta G}_{298}^E$ (cal mol^{-1})	$\overline{\Delta H}_{298}^E$ (cal mol^{-1})	$\overline{\Delta S}_{298}^E$ (cal mol^{-1} K^{-1})
Gel phase	4430	14	258	65	1107	3.5
Liquid-crystalline phase	500	1	202	50	125	0.25

[a] Blume (unpublished).

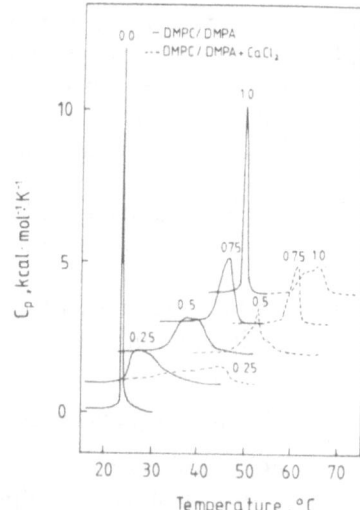

Figure 10. DSC scans of DMPC–DMPA mixtures with and without Ca^{2+}; Numbers designate mole fraction of DMPA.

temperature to ca. 61°C. For mixtures with DMPC similar changes in T_m are observed, but the shapes of the DSC curves are altered drastically. The uncorrected experimental phase diagram attains a form indicating strongly nonideal mixing behavior in the gel phase on the right-hand side of the phase diagram (see Fig. 11). The phase diagram may even be of the peritectic

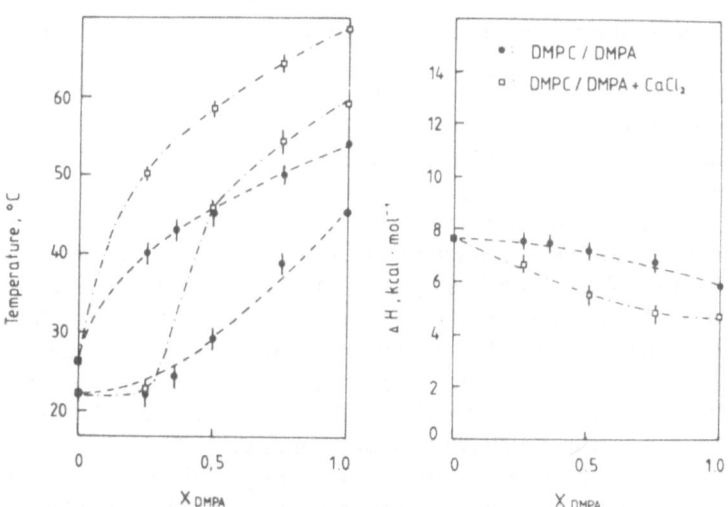

Figure 11. Left: Phase diagram for DMPC–DMPA mixtures; ●, onset and end of melting determined from the DSC scans in Fig. 10 for mixtures without Ca^{2+}; □, onset and end of melting for mixtures with Ca^{2+}. Right: ΔH versus mole fraction DMPA for DMPC–DMPA mixtures with and without Ca^{2+}.

type with a miscibility gap in the gel phase. Though the liquid-crystalline phase still seems to be homogeneous, the strong ideality will lead to formation of large clusters of PA molecules. This effect has indeed been experimentally observed before using spectroscopic techniques (Ito and Ohnishi, 1974; Galla and Sackmann, 1975). This example was shown to illustrate that changes in ion concentrations can result in drastic changes of the mixing behavior, which will lead to a different lateral organization of the bilayer. This phenomenon is interesting because for biological membranes it might have some functional implications. It offers the possibility of changing the activity of an enzyme in the membrane by an increase or decrease in ion concentration in an indirect fashion (Träuble, 1976).

2.3.4. Phospholipid/Cholesterol Mixtures

Cholesterol is present in the membranes of most eukaryotic cells. The thermotropic behavior of lipid-cholesterol mixtures has therefore been under intensive investigation (Demel and de Kruyff, 1976). Ladbrooke *et al.* (1968) have shown that addition of cholesterol to DPPC causes a disappearance of the pretransition, a broadening of the DSC peak, and finally a complete disappearance of the endothermic main transition when the cholesterol concentration reaches 50 mol%. These results were later confirmed with more sensitive DSC instruments (Estep *et al.*, 1984; Mabrey *et al.*, 1978) and it was shown that below a concentration of 20 mol% cholesterol the DSC peaks can be resolved into two component peaks, a narrow one at almost the same temperature of the pure DPPC and a broad one shifted to higher temperature. Above 20 mol% cholesterol, only the broad peak remains, vanishing finally at 50 mol% cholesterol. Cholesterol has a similar effect on the thermotropic properties of other phospholipids like sphingomyelins, phosphatidylethanolamines, phosphatidic acids, and phosphatidylglycerols (Estep *et al.*, 1979; Calhoun and Shipley, 1979; Blume, 1980; Blume and Hillmann, 1986), though the shape of the calorimetric endotherms at intermediate cholesterol concentrations depends on the nature of the lipid head group and on the fatty acid chain length. For DMPC, for instance, cholesterol causes an increase in T_m (Mabrey *et al.*, 1978), whereas for DMPE and DMPA cholesterol incorporation reduces T_m (Blume, 1980; Blume and Hillmann, 1986). Figures 12 and 13 show DSC scans of DMPE-cholesterol and DMPA-cholesterol mixtures to illustrate the thermotropic behavior of these mixtures. The effects caused by cholesterol incorporation arise from the rigidity of the steroid ring system. Cholesterol acts as a "spacer" in lipid bilayers, separating the head groups and thus reducing specific interactions in the case of PE and PA. But it also lowers the van der Waals interaction energies between the chains, thus reducing the transition enthalpy (Demel and de Kruyff, 1976). In the

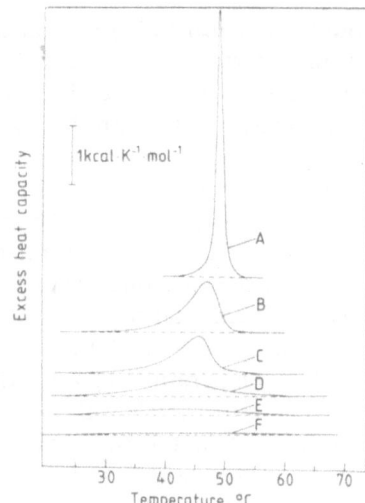

Figure 12. DSC scans for DMPE–cholesterol mixtures. A, 0 mol%; B, 5 mol%; C, 10 mol%; D, 20 mol%; E, 30 mol%; and F, 40 mol% cholesterol (Blume, 1980).

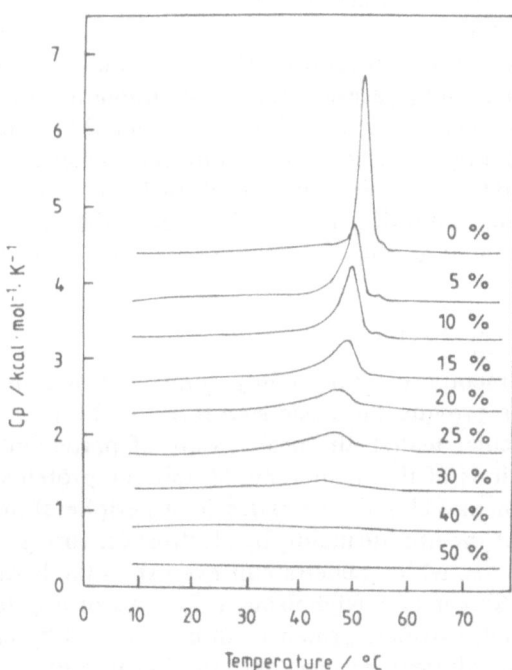

Figure 13. DSC scans for DMPA–cholesterol mixtures. Numbers designate mol% cholesterol (Blume and Hillmann, 1986).

liquid-crystalline phase the mean molecular area of the lipids at the bilayer-water interface is reduced. This is the so-called condensation effect of cholesterol. *Trans-gauche* isomerization of the chains is also hindered, particularly in the middle part of the acyl chains next to the rigid ring system. Thus cholesterol transfers the lipid molecules into some intermediate state with respect to the gel and liquid-crystalline phase.

Numerous attempts have been made to interpret the thermotropic behavior of lipid–cholesterol mixtures in terms of a phase diagram, and very complicated ones have been suggested on the basis of investigations using a combination of different techniques (Lentz *et al.*, 1980). Phase boundaries between 20 and 25 mol% cholesterol have been proposed for the low-temperature region of the phase diagram leading to phase separation into cholesterol-rich and cholesterol-free domains. NMR and X-ray studies, on the other hand, could not confirm these results. The behavior of lipid–cholesterol mixtures is thus not completely understood.

An alternative explanation based on theoretical arguments by Jähnig (1981) and supported by experimental evidence from X-ray (Hui and He, 1983), NMR (Blume and Griffin, 1982), and pressure jump relaxation studies (Blume and Hillmann, 1986) is that the phase transition changes its character from first order at lower cholesterol content to second order at some critical percentage, which in turn depends on the nature of the lipid. Thus at higher cholesterol content the transition would be continuous, and because of its second-order character phase separation into domains of different physical state could not occur. For a biological system this would be of great advantage as cooling of the membrane into the transition region would in this case not lead to phase separation resulting in an aggregation of proteins with concomitant inhibition of function. In addition, leaks, which are preferentially located at phase boundaries, are also reduced.

2.3.5. Lipid/Protein Mixtures

Biological membranes are mainly composed of lipids and proteins. While the lipids provide the basic structure for the permeability barrier, i.e., the bimolecular leaflet, the proteins are of particular importance for the active functions of the membrane. Membrane proteins can be divided into two fundamental classes, the extrinsic or peripheral proteins bound to the outside of the membrane mainly by electrostatic forces, and the intrinsic or integral proteins, which penetrate or even span the hydrophobic region of the bilayer (Singer and Nicholson, 1972). According to their different modes of binding, extrinsic proteins can be removed by changing the pH or ionic strength, whereas intrinsic proteins, because of the hydrophobicity of some parts of their polypeptide chain, can only be isolated by extracting the membrane with detergents. As most intrinsic proteins have also hydrophilic regions exposed to water, electrostatic interactions with peripheral proteins or with lipids surrounding the protein are also possible.

Lipid–protein interactions have been studied intensively over the past years using a variety of techniques (Papahadjopoulos, 1974). One of the major questions has been, in what state are the lipids adjacent to the protein, called the lipid annulus? Spectroscopic techniques can give information on the dynamical and conformational properties of these molecules (Seelig, 1982; Marsh and Watts, 1982). Calorimetric measurements enable us to gain insight into their thermodynamic properties.

For the characterization of lipid–protein interactions DSC measurements are of great help as this method is fast and very sensitive so that only little material is needed. This is particularly important in cases where only limited amounts of the membrane protein under investigation can be isolated and reconstituted. In addition DSC is a nonperturbing technique obviating the necessity to add a probe molecule, so that the sample can be used for other types of studies after the DSC measurements without having to go through a repurification procedure.

In the DSC experiments the shape of the transition peak of the phospholipid is studied as a function of protein content. According to the effects the intrinsic proteins have on the shape of the DSC peak they can be divided into different classes. Either protein incorporation decreases the transition temperature or not. Sometimes even an increase in T_m is observed. In all cases the transition enthalpy ΔH is lowered with increasing protein content and in most cases the transition is also broadened. Proteins have perturbing effects on the lipids in the annulus surrounding the protein. Cooperative interactions are reduced because the lipids are in some thermodynamically intermediate state and have a reduced number of other lipid molecules as neighbors. A plot of the transition enthalpy ΔH versus the protein/lipid molar ratio can give information on the stoichiometry of the lipid–protein interaction. In simple cases when this plot gives a straight line the number of lipids removed from the transition can be determined and then identified with the number of lipid molecules in the first "solvation shell." One has to assume, however, that lipids in the annulus do not take part in the transition. In many cases the situation is complicated by protein aggregation at low temperatures when the lipids are in the gel state. Lipid molecules can then also be trapped in between the proteins and thus removed from the transition without necessarily belonging to the annulus.

A well-defined example of a lipid–protein system where protein aggregation occurs at low temperature in the gel state is the bacteriorhodopsin (BR)–DMPC mixture (Heyn et al., 1981a). Figure 14 shows DSC scans of BR–DMPC vesicles with several different lipid/protein ratios. Table V summarizes the calorimetric data. A plot of ΔH versus the protein/lipid ratio would give a number of ca. 39 molecules of lipid removed from the

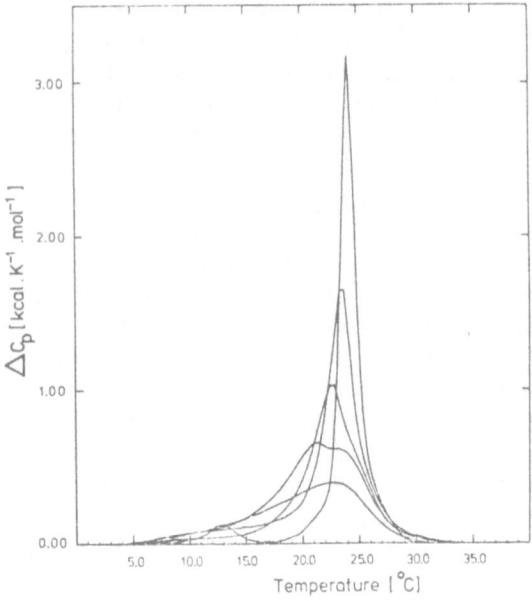

Figure 14. DSC scans of DMPC–BR vesicles with the following ratios: ∞, 552, 316, 188, 91 (Heyn *et al.*, 1981a).

transition. In the case of BR, however, the temperature-dependent aggregation and the formation of the BR lattice at low temperature can be studied directly by other techniques, such as CD measurements (Heyn *et al.*, 1981b). From the latter experiments it was known that BR aggregates well below the DMPC transition with an aggregation midpoint of 17.5°C. From the slope of the CD transition curve the van't Hoff transition enthalpy for the formation of the protein lattice was calculated to be 55 kcal mol^{-1} (Heyn *et al.*, 1981b). In the course of the dissociation of the protein lattice coming from low temperatures additional lipids are required as "solvation lipids."

Table V. Calorimetric Data for DMPC/BR Mixtures[a]

DMPC/BR ratio	T_m (°C)	ΔH_{cal} (kcal mol^{-1})	$\Delta H_{v.H.}$ (kcal mol^{-1})
∞	24.1	7.6	287
552	23.3	7.3	134
316	23.2	6.0	113
188	23.0	6.3	68
91	22.8	4.4	55

[a] Heyn *et al.* (1981a).

Under the assumption that these lipids are in some intermediate thermodynamic state, i.e., that they are energetically lower than pure liquid-crystalline lipid but higher than gel state lipid, the transition curves derived from the DSC experiments can be simulated using the following equation:

$$\theta(T) = \frac{1}{\Delta H_{\text{cal}}}\left[\frac{\text{BR}}{L}\, m\, \Delta H_{1_{\text{cal}}}\theta_1(T) + \left(1 - \frac{\text{BR}}{L}\right)\Delta H_{2_{\text{cal}}}\theta_2(T)\right] \quad (26)$$

with m the number of "solvation lipids," $\Delta H_{1_{\text{cal}}}$ the enthalpy change for the "solvation lipids," $\Delta H_{2_{\text{cal}}}$ the transition enthalpy for bulk lipids, and BR/L the bacteriorhodopsin/DMPC ratio. ΔH_{cal} is given by

$$\Delta H_{\text{cal}} = (\text{BR}/L)_m \Delta H_{1_{\text{cal}}} + (1 - (\text{BR}/L)m)\Delta H_{2_{\text{cal}}} \quad (27)$$

$\theta_1(T)$ and $\theta_2(T)$ are the degrees of transition for the solvation lipids melting at 17.5°C and bulk lipid melting at a higher temperature T close to the transition temperature of pure DMPC. $\theta_1(T)$ and $\theta_2(T)$ are given by

$$\theta_1(T) = 1 \bigg/ \left\{1 + \exp\left[\frac{-\Delta H_{\text{v.H}_1}}{R}\left(\frac{1}{T} - \frac{1}{T_1}\right)\right]\right\} \quad (28)$$

$$\theta_2(T) = 1 \bigg/ \left\{1 + \exp\left[\frac{-\Delta H_{\text{v.H}_2}}{R}\left(\frac{1}{T} - \frac{1}{T_2}\right)\right]\right\} \quad (29)$$

with $\Delta H_{\text{v.H}_1} = 55$ kcal mol^{-1} and $\Delta H_{\text{v.H}_2}$ being the van't Hoff enthalpies of melting for solvation and bulk lipids, respectively (Heyn *et al.*, 1981a).

Figure 15 shows the experimental transition curves and the dotted lines are the calculated curves according to the model described above. The parameters used for the simulation are summarized in Table VI. From this model calculation a number of $m = 60$ is deduced for the solvation lipids. This would mean that more than one shell of lipids surrounding the bacteriorhodopsin molecule is perturbed, as only 30 molecules are estimated

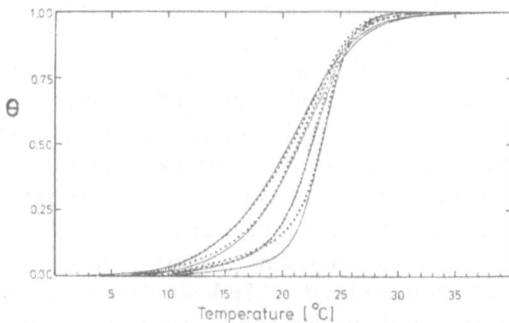

Figure 15. Solid line; normalized transition curves calculated from the DSC scans in Fig. 14; dotted line: calculated transition curves using the model described in the text (Heyn *et al.*, 1981a).

Table VI. Parameters Used for the Simulation of the Calorimetric Transition Curves for DMPC-BR Mixtures[a]

DMPC/BR ratio	T_1 (°C)	$\Delta H_{\mathrm{v.H._1}}$ (kcal mol^{-1})	T_2 (°C)	$\Delta H_{\mathrm{v.H._2}}$ (kcal mol^{-1})	Solvation lipids (M per BR)	$F = \Delta H_{2_{\mathrm{cal}}}/\Delta H_{1_{\mathrm{cal}}}$
552	17.5	55	23.5	150	60	0.7
316	17.5	55	22.9	115	60	0.7
188	17.5	55	22.3	80	60	0.7
91	17.5	55	22.3	80	50	0.7

[a] Heyn *et al* (1981a).

from the dimensions of the protein to form the first layer. The second important result is that the enthalpy change for the "solvation lipids" $\Delta H_{2_{\mathrm{cal}}}$ is only 70% of $\Delta H_{1_{\mathrm{cal}}}$, the transition enthalpy of free lipid. As mentioned above, this indicates that lipids surrounding the BR molecule are in some lower enthalpic state as compared to free lipids. This could be interpreted as an "ordering effect" of the protein. We will see that these results obtained for the BR–DMPC system by DSC can be compared well with the enthalpy change measured directly by reaction calorimetry for the phospholipid–melittin system (see below).

As a second example of the applications of DSC to lipid–protein mixtures we want to discuss the thermotropic behavior of a system where the protein is a small amphiphilic polypeptide which is water soluble but partitions also into the lipid bilayer. The 26 amino acid long polypeptide melittin from bee venom forms a bent α helix with a hydrophobic and a hydrophilic side and a very polar N-terminus due to four positively charged amino acids (Habermann, 1972; Terwilliger, 1982). In aqueous solution melittin aggregates into tetramers where the hydrophobic surfaces of the α helices are hidden from water (Habermann, 1972; Terwilliger, 1982). In the presence of lipid bilayers melittin partitions also into the membrane. However, it is not perfectly clear how the molecule is oriented in the bilayer.

DSC studies of a variety of phospholipid–melittin mixtures have been performed. In all cases the transition enthalpy for the lipid was decreased (Mollay, 1976; Bernard *et al.*, 1982). As an example we want to show DSC scans of DMPC–melittin mixtures (see Fig. 16). The decrease of the peak height with increasing melittin concentration is obvious, but T_m and the width of the phase transition seems to stay almost invariant. With the negatively charged phospholipid DMPA similar results are obtained (see Fig. 17). In this case a slight increase in T_m seems to occur. The dependence of the transition enthalpy on melittin content is shown in Fig. 18. The plots are not linear as in other lipid–protein systems but display a curvature,

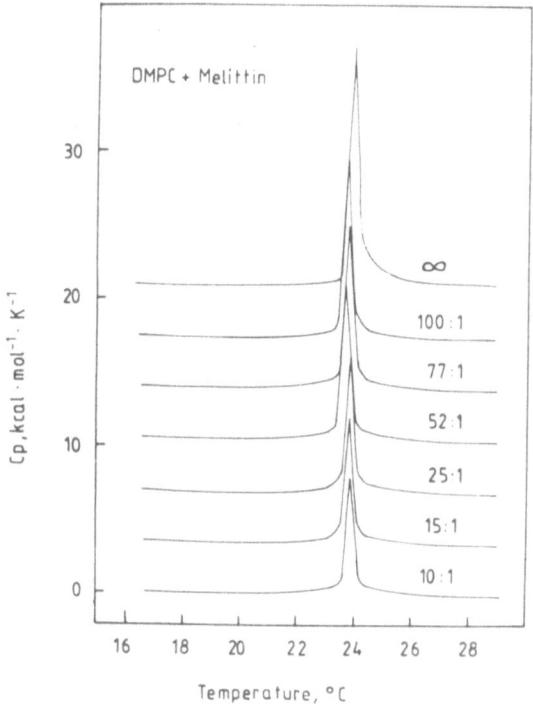

Figure 16.   DSC scans of DMPC–melittin mixtures. Numbers designate DMPC/melittin ratio.

which is particularly strong for DMPC. This indicates that at low concentrations of melittin the number of lipids perturbed or "removed" from the transition is higher. A possible cause for this phenomenon is the aggregation of melittin in the bilayer to oligomers, possibly to tetramers (Vogel *et al.*, 1983; Gevod and Birdi, 1984). When melittin is incorporated into DMPG or DPPG bilayers an additional electrostatic effect can be observed. Figure 19 shows DSC curves of various DPPG–melittin mixtures and Fig. 20 the ΔH dependence. ΔH decreases in a nonlinear fashion as in the other lipid–melittin systems, but in this case the transition is broadened and shifted to higher temperature. This is probably caused by electrostatic interactions between the four positively charged amino acids at the end of the melittin α helix with the negatively charged phosphate groups of DPPG. Divalent cations like Ca^{2+} increase T_m of DPPG to 58°C (see Fig. 28). In the case of melittin T_m is increased to 52°C at high protein content, so the electrostatic effect is smaller than for Ca^{2+}. For the other negatively charged lipid DMPA this electrostatic effect is masked by the strong intermolecular interactions between the PA head groups via hydrogen bonds. When these interactions are reduced, as in DMPC–DMPA mixtures due to the "spacer

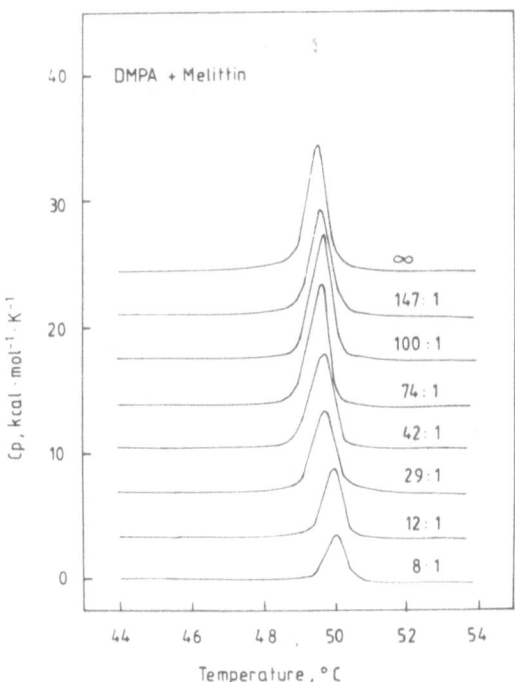

Figure 17. DSC scans of DMPA-melittin mixtures. Numbers designate DMPA/melittin ratio.

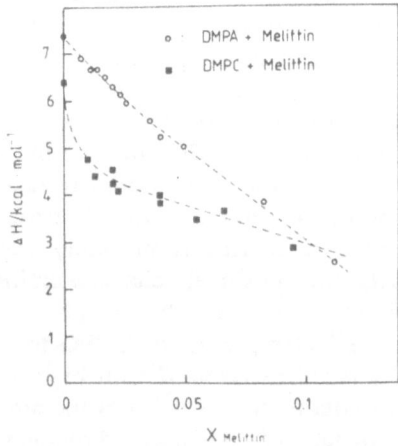

Figure 18. Transition enthalpy ΔH for DMPC-melittin and DMPA-melittin mixtures as a function of mole fraction melittin.

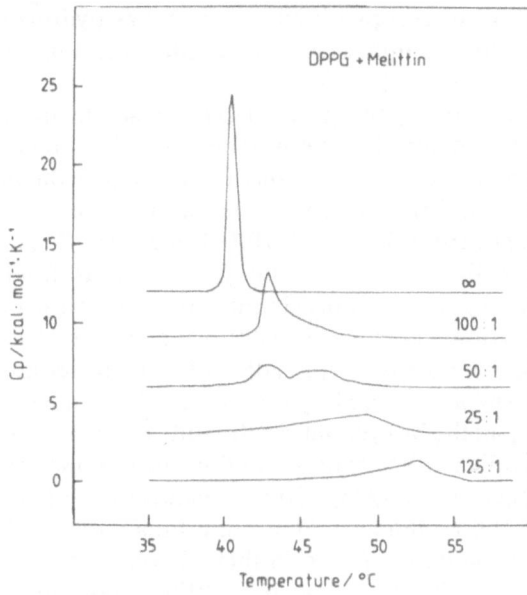

Figure 19. DSC scans of DPPG–melittin mixtures. Numbers designate DPPG/melittin ratio.

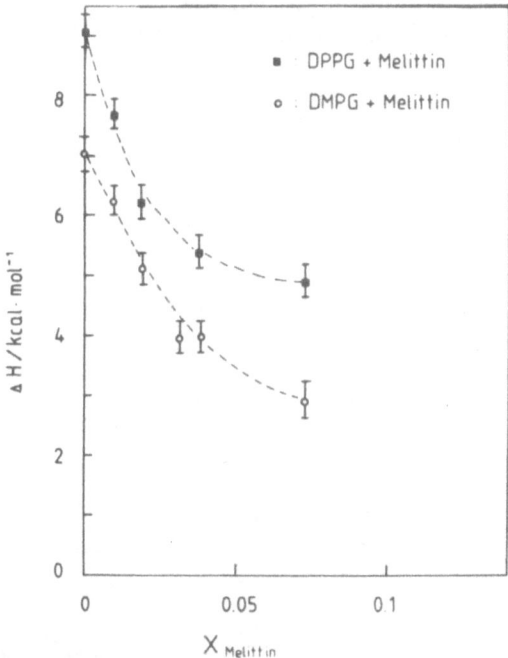

Figure 20. Transition enthalpy ΔH for DMPG–melittin and DPPG–melittin mixtures as a function of mole fraction melittin.

effect" of the PC head groups which cannot act as hydrogen bond donors, the electrostatic effect is increased. This is shown in Fig. 21 for a DMPC–DMPA (1:2.3) mixture.

As mentioned above, the plots of ΔH versus melittin concentration can be analyzed to determine the number of lipids "removed" from the transition, a number that can be loosely interpreted as the lipids in the first solvation shell around the protein. At melittin concentrations of 5 mol% this number ranges from 6.5 for DMPA, 8 for DMPC, to 10 for DMPG. For a simple α helix spanning the bilayer, ca. 13–14 lipid molecules are expected to be in the first annulus simply from geometrical arguments, i.e., from the cross-sectional area of the melittin α helix, which is approximately 80 Å^2. There are two possible explanations for the lower numbers found in the experiment. The first possibility is that the orientation of melittin is such that it does not span the bilayer but resides only in one half of the membrane with the hydrophilic surface exposed to the surrounding water (Terwilliger *et al.*, 1982; Vogel *et al.*, 1983). From a calculation of the length of the melittin α helix and its thickness one can estimate that in this case at least 8–10 molecules should be affected in their thermal behavior.

The second possibility is that the melittin molecules aggregate into tetramers where the hydrophilic surfaces of the helices are oriented away from the apolar chains of the lipids. This form of aggregation was proposed to act as an ion channel and could also be observed in experiments using

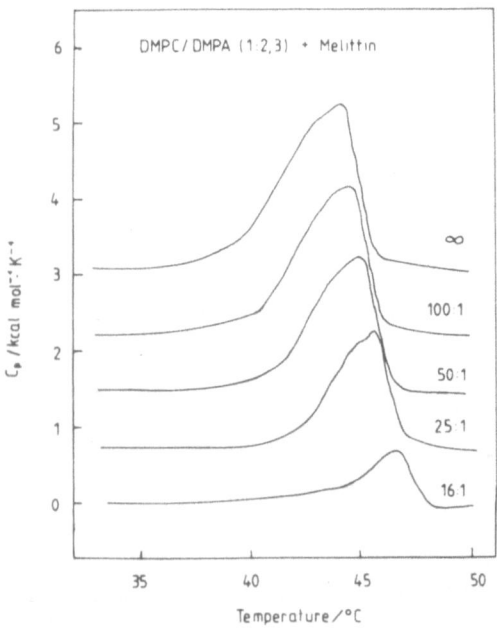

Figure 21. DSC scans of DMPC–DMPA–melittin mixtures with lipid/melittin ratios as indicated.

lipid–melittin monolayers (Gevod and Birdi, 1984; Tosteson and Tosteson, 1981). In this case ca. 22–23 molecules would be in the first solvation shell of the tetramer; that means a number of ca. 5–6 per melittin monomer. The numbers from the DSC experiments are between 6 and 10 at a lipid/protein ratio of 20:1. However, the ΔH plots are curved, and at lower protein content the numbers become larger. Because at lower total protein content the monomer concentration should be increased relative to the number of tetramers, the observed curved ΔH plots seem to support the monomer–tetramer model with a concentration-dependent degree of aggregation. An alternative explanation like a concentration-dependent change of orientation cannot be excluded, however.

The limitations of the DSC method in these types of studies of lipid–protein interactions are evident. The measurements are rather indirect as only the lipids, i.e., their temperature-dependent behavior, are observed. Nevertheless the characterization of lipid–protein mixtures by DSC is very valuable as first information is obtained on protein aggregation, on electrostatic effects, and on the number of lipids in the lipid annulus being perturbed by the protein. In addition, in high-sensitivity DSC relatively small amounts of material are needed and the experiments are simple and fast to perform.

3. REACTION CALORIMETRY

3.1. Principle and Instrumentation

The principle of reaction calorimetry is very simple. Two reactants are mixed together in a reaction cell and the heat evolved or needed is measured in some way. In the classical combustion calorimeter this is done by determining the temperature change of a water bath surrounding the combustion bomb. There are numerous other ways of determining the heat evolved during a reaction. In many cases the temperature difference between the reaction cell and the thermostatted surrounding is detected by thermopiles and the electrical signal coming from the thermopiles is directly recorded as a function of time on an x-t recorder. The calibration of the apparatus is usually performed by applying a precisely known current through a calibration resistor in the reaction cell for a predetermined time so that an exactly known amount of heat is evolved.

For applications to biological or biochemical problems several reaction calorimeters, manufactured by Tronac (Orem, Utah, USA), Setaram Instruments (Lyon, France), or LKB Instruments (Bromma, Sweden), to name only a few, are commercially available. Here, only the LKB Model 10700 Batch Microcalorimeter with titration accessory will be described. A schematic drawing of the cell assembly of this instrument is shown in Fig. 22.

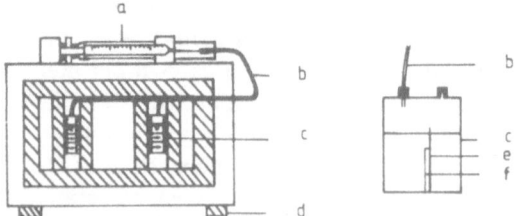

Figure 22. Schematic drawing of the LKB 10700 Microcalorimeter cell assembly with titration accessory and cross section through one of the reaction cells; (a) syringes driven by step motor; (b) Teflon capillary tubing; (c) reaction cell; (d) counter weights of the drum of the cell assembly; (e) partitioning wall; (f) calibration heater; after Chen and Wadsö (1982).

This calorimeter can be used as a batch-type instrument, where both reactants are placed in the two compartments of cell 1. One compartment of the reference cell 2 is then filled with reactant A, the other, for instance, with buffer. In this way heats of dilution can be compensated. Alternatively, this instrument can also be used as a titration calorimeter (Chen and Wadsö, 1982). In this type of experiment reactant A is placed in cell 1 and cell 2 contains only buffer. Reactant B is then filled into both syringes and is then titrated into both cells. The reaction with A occurs in cell 1; the heat of dilution of B into buffer occurs in the reference cell 2 and is compensated. In both types of experiments mixing of the reactants is achieved by turning the cell assembly several times upside down and back again to the original position. For reactions in solution using the titration mode heat quantities of ca. 1–2 mcal can be determined with sufficient accuracy (Chen and Wadsö, 1982). In applications using phospholipid bilayers the accuracy is generally somewhat lower because these systems are heterogeneous and the reactions are sometimes quite slow. Larger heats of reaction are then required to improve the accuracy.

3.2. Determination of the Heat of Ion Binding to Phospholipids

Reaction calorimetry can be used to study heats of ion binding to charged lipids like PA, PG, or PS. Phospholipids with dissociable protons in the head group have different T_m and ΔH values depending on the head group charge (Blume, 1983; Träuble et al., 1976). DMPA, for instance, in its singly charged form, has a T_m value of ca. 52°C with a transition enthalpy of 5.7 kcal/mol. At pH 12, the phosphate group is doubly charged and T_m decreases to 22.4°C and ΔH to 4.1 kcal/mol (Blume, 1983; Blume and Eibl, 1979). These differences in thermotropic properties are an indication of changes in bilayer structure. Indeed, Jähnig et al. (1979) found that the angle of tilt of the hydrocarbon chains of a phosphatidic acid increases with head group charge.

The deprotonation of the PA head group can be studied by reaction calorimetry by mixing a DMPA dispersion at pH 7, where the head group is singly charged, with a 10^{-2} N NaOH solution. This leads to a pH change and a complete deprotonation of the PA head group. The overall reaction is

$$DMPA^- + OH^- \rightleftharpoons DMPA^{2-} + H_2O \qquad (30)$$

The heat of reaction will be the sum of the heat of dissociation ΔH_{diss} of DMPA, including a possible reaction heat caused by a structural change in the bilayer, plus the heat of neutralization ΔH_{neutr}. The heat of dilution of NaOH can be compensated in the reference cell. Values for the heat of neutralization ΔH_{neutr} can be found in the literature or determined separately by mixing HCl with NaOH. The heat of dissociation ΔH_{diss} is then obtained by subtracting ΔH_{neutr} from the total heat of reaction. ΔH_{diss} and ΔH_{neutr} are of course temperature dependent. Figure 23 shows the ΔH_{diss} values for the dissociation of the second proton of DMPA as a function of temperature in the experimentally accessible range between 0 and 60°C.

This diagram can be divided into three different regions: (a) below 22°C the pH jump induces a deprotonation plus a change of the tilt angle of the acyl chains, so that ΔH_{diss} includes a contribution from the change of structure from an L_β to an $L_{\beta'}$ phase; (b) between 22 and 51°C deprotonation will induce a phase change from L_β to L_α, so that ΔH_{diss} includes the transition enthalpy from L_β to L_α; above 52°C ΔH_{diss} will be approximately equivalent to the true heat of dissociation as the character of the

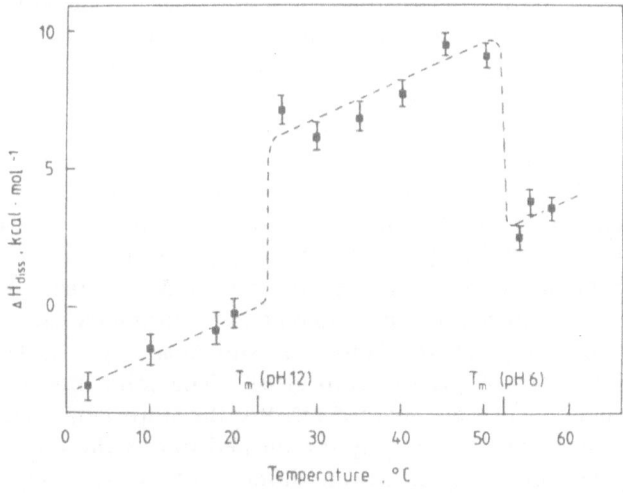

Figure 23. Temperature dependence of the dissociation enthalpy ΔH_{diss} for the second proton of DMPA.

lamellar L_α phase does not change measurably. As can be seen from Fig.
23 ΔH_{diss} is not constant within these three different regions. The slope of
the ΔH vs. T plot gives directly the value for the change in heat capacity
of the system during the reaction: $\Delta C_p = (d\Delta H/dT)$. In our case it is the
difference in heat capacities between singly and doubly charged DMPA.
The slope in the plots is positive and we get a value of ca. 140 cal K^{-1} mol^{-1}
for ΔC_p. Data for the heat of dissociation of a water soluble molecule
resembling only the head group of DMPA, the compound glycero-1-phos-
phate can be taken from the literature and compared to the values for
DMPA. ΔH_{diss} for glycero-1-phosphate is also strongly temperature dependent, but in the opposite way, i.e., ΔC_p is negative with ca. -55 cal K^{-1} mol^{-1}
(Datta and Grzybowski, 1958). As mentioned in Section 1.3.2, the apparent
molar heat capacities $^\phi C_p$ include contributions from hydrophilic as well
as "hydrophobic hydration." For glycero-1-phosphate the increase in charge
upon dissociation leads to a characteristic decrease of $^\phi C_p$, observed
similarly for most dissociation reactions where additional charges are generated. The opposite sign of ΔC_p for DMPA, i.e., the increase in $^\phi C_p$ for the
doubly charged form of the lipid must arise from an overcompensation of
the expected decrease of the heat capacity by other effects. It seems likely
that this arises from an increase in "hydrophobic hydration" induced by
the change in structure, namely, that the repulsive electrostatic effects lead
to larger separation of the lipid molecules in the bilayer facilitating the
access of water molecules to hydrophobic regions of the bilayer. A second
contribution to the positive ΔC_p could arise from an increase in counterion
condensation on the negatively charged bilayer surface. As water of hydration can then be shared between the counter ions and the negative phosphate
groups free water will be formed having a higher heat capacity. So the sum
of these two effects seems to overcompensate the normally negative ΔC_p
found for dissociation reactions.

These results obtained by reaction calorimetry can also be checked by
DSC measurements as described in Section 1.3.2. One has to perform a
DSC scan of DMPA at pH 7 and compare it with a DSC scan of DMPA
at pH 12. The difference in apparent molar heat capacity $^\phi C_p$ determined
this way is ca. 150 cal K^{-1} mol^{-1} and compares well with the ΔC_p value
obtained from the temperature dependence of ΔH_{diss} (see above). Data
derived from DSC measurements and reaction calorimetry can be combined
in an enthalpy-temperature diagram as shown in Fig. 24. The enthalpy
function H is arbitrarily set to zero at 0°C. The solid line represents the
enthalpy function of singly charged DMPA, the slope being determined by
$^\phi C_p$ measured by DSC. Similarly the dashed line is the enthalpy for the
doubly charged form. The slopes are different of course as DMPA^{2-} has a
higher $^\phi C_p$ value. The fat arrows represent the transition enthalpies ΔH
measured for both forms by DSC and the small arrows the heats of

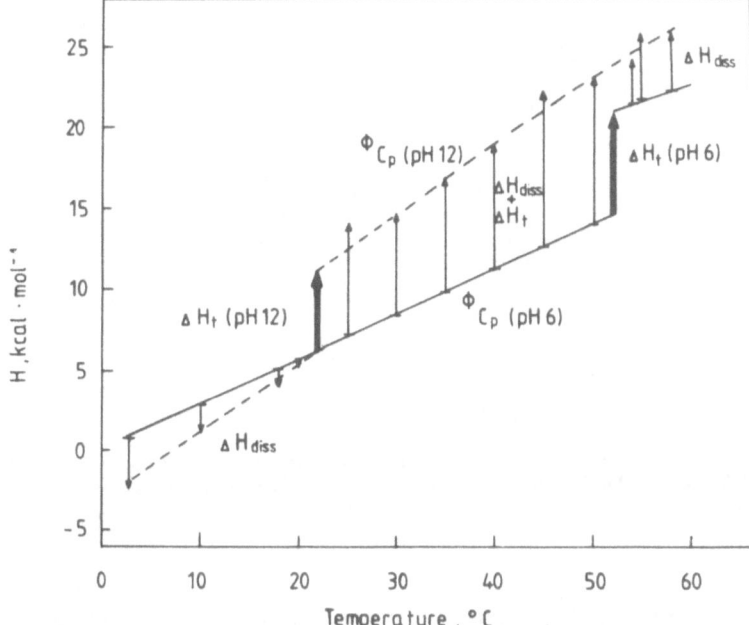

Figure 24. Enthalpy-temperature diagram for the $DMPA^-$–$DMPA^{2-}$ system. Solid line: enthalpy versus temperature for $DMPA^-$; fat arrow at 52°C designates transition enthalpy for gel to liquid-crystalline phase transition; the slope is determined by $^\phi C_p$ as measured by DSC; dashed line: enthalpy versus temperature for $DMPA^{2-}$; fat arrow at 22°C marks the transition enthalpy for the gel to liquid-crystalline phase transition; small arrows are dissociation enthalpies as measured by reaction calorimetry (see Fig. 23).

dissociation ΔH_{diss} measured at fixed temperatures by reaction calorimetry. The lengths of the arrows correspond to the ΔH_{diss} data in Fig. 23. So both data sets fit nicely into one diagram, giving a complete thermodynamic description of the DMPC–DMPA system.

The dissociation reaction of DMPA can also be followed by a tritration experiment using the titration accessory described in Section 2.1. For this experiment an exactly measured amount of a DMPA dispersion of known concentration at pH 7 is placed in the reaction cell and titrated with 0.5 N NaOH solution. The heat of dilution is again compensated in the reference cell by titrating the NaOH solution into buffer. The integral heat of reaction as a function of added NaOH solution for the titration of DMPA is shown in Fig. 25. The shape of the curve resembles a binding isotherm. However, the binding constant, which in our case is the reciprocal value of the dissociation constant, is only an apparent constant as it depends on the surface potential ψ_0 (Träuble, 1976):

$$K_{\text{app}} = K_0 \exp(e\psi_0 / kT) \tag{31}$$

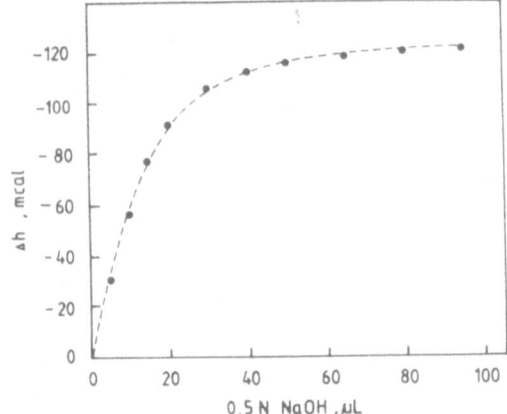

Figure 25. Integral heat of reaction for a titration experiment using 5 ml 1.25×10^{-3} M DMPA and 0.5 M NaOH. Dashed line: titration curve calculated using equation (32) and intrinsic pK of 6.2.

with the surface potential ψ_0 being proportional to the logarithm of the square of the surface charge density σ. From the apparent dissociation constant the degree of dissociation α can be determined. It also depends on the surface potential according to the following equation (Träuble, 1976):

$$\alpha = K_0/\{K_0 + [H^+]\exp(e\psi_0/kT)\} \tag{32}$$

with $[H^+]$ being the aqueous proton concentration. For a lipid bilayer the surface charge density can be calculated from the surface area f occupied by a lipid and the number of charges per head group. For DMPA between pH 7 and 12 the electric charge changes from -1 to -2. Thus the surface charge density σ is

$$\sigma = (1 + \alpha)e/f \tag{33}$$

with e being the electric charge. Approximating ψ_0 for high surface charge densities and inserting the appropriate values for e, f, k, and T (Träuble, 1976) leads to the following equation for the dissociation of the second proton of DMPA:

$$[H^+] = \frac{(1 - \alpha)K_0 n_m}{\alpha(1 + \alpha)^2 \cdot 29.15} \tag{34}$$

with n_m the monovalent salt concentration of the solution in mol l^{-1}. With the help of equation (34) the titration curve of Fig. 25 can be calculated under the assumption that the integral heat of reaction Δh is directly proportional to the degree of dissociation α and with K_0 as an adjustable parameter. The dashed line in Fig. 25 has been calculated with $K_0 = 10^{-6.2} \text{ mol } l^{-1}$. This value is almost the same as the dissociation constant of the water soluble compound glycero-1-phosphate. Thus surface charge effects are responsible for the large shift of the apparent dissociation constants for DMPA to pK values of ca. 10, and it is clear from equation (31) that the apparent dissociation constant changes with the degree of dissociation.

It was mentioned in Section 1.3.3 that negatively charged lipids like DMPA strongly bind divalent cations like Mg^{2+} or Ca^{2+}. The binding of Mg^{2+} or Ca^{2+} to DMPA shifts the transition temperature to ca. 60–65°C and decreases the transition enthalpy (see Fig. 10) (Träuble, 1976; Elamrani and Blume, 1984). The binding of Ca^{2+} to DMPA can also be followed by a titration experiment in the LKB reaction calorimeter. Figure 26 shows the integral heat of reaction when singly charged DMPA is titrated with $CaCl_2$ solution. The curve goes into the saturation limit at a Ca^{2+} to DMPA$^-$ ratio of 0.5, clearly indicating stoichiometric binding of Ca^{2+} under these experimental conditions. As in the case of dissociation of DMPA the apparent binding constant depends on the surface charge density, in this case decreasing with increasing degree of binding. The dashed curve in Fig. 26 is calculated using a similar expression to equation (34) (Woolley and Teubner, 1979). The calculated intrinsic binding constant K_0 for Ca^{2+} binding used in the simulation is ca. 7 M^{-1}. This value is somewhat higher than K_0 for Ca^{2+} binding to methylphosphatidic acid (Woolley and Teubner, 1979) or Ca^{2+} binding to phosphatidylserine monolayers (Ohki and Sauve, 1978). The value of K_0 would give an apparent binding constant K_{app} of

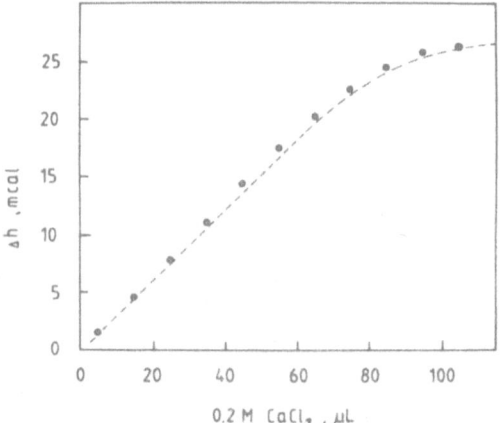

Figure 26. Integral heat of reaction for a titration of 5 ml 0.01 M DMPA with 0.2 M CaCl. Dashed line: calculated titration curve for 1:2 binding with an intrinsic binding constant K_0 of 7 M^{-1}.

ca. $6 \times 10^6 \, M^{-1}$ at the particular salt concentrations we used in our experiments. This compares well with other known values for K_{app} for binding of Mn^{2+} to phosphatidylglycerol (Puskin and Martin, 1979) or Ca^{2+} to phosphatidylserine (Hauser *et al.*, 1976). In reality the binding of divalent cations in the presence of monovalent cations can be more complex than assumed for the model calculations of the binding curve in Fig. 26. In particular 1:2 binding can produce excluded sites not accessible to further Ca^{2+} ions, 1:1 binding of Ca^{2+} may also occur, and in addition monovalent cation binding may be competitive (Cohen and Cohen, 1981). The relatively good fit we obtained with the simple model, however, makes it questionable whether more complicated models are really needed, at least in this particular case of binding of Ca^{2+} to DMPA.

If we turn to the total binding enthalpy of Ca^{2+} to DMPA$^-$ we find that at low temperature ΔH is positive. This is also true for Mg^{2+} binding. Increasing the temperature leads to a decrease of ΔH till it changes sign at ca. 30–35°C (see Fig. 27). The ΔC_p accompanying the binding reaction is therefore negative and is determined from the slope of the ΔH vs. T plot of Fig. 27 to $-20 \, \text{cal K}^{-1} \, \text{mol}^{-1}$. This negative ΔC_p can be contrasted with results obtained for Mg^{2+} binding to mononucleotides and polynucleotides. In these cases Mg^{2+} binding is connected with a positive ΔC_p of 8–16 cal K^{-1} mol^{-1} arising from the liberation of water of hydration as the

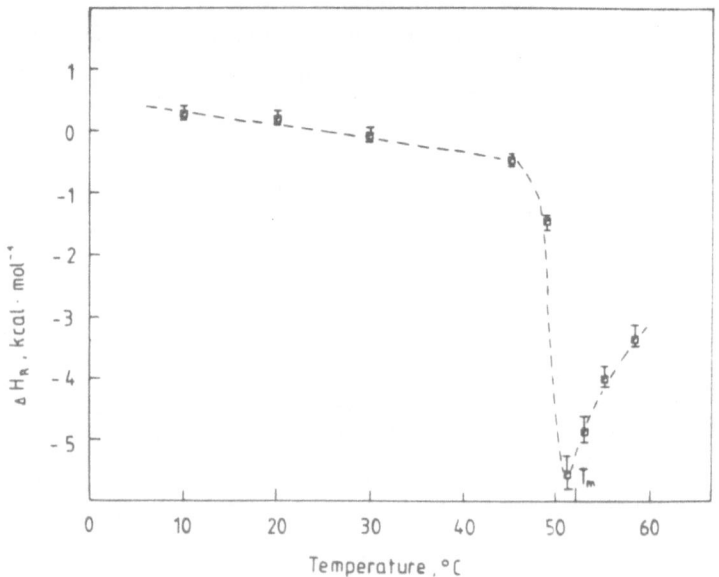

Figure 27. Temperature dependence of the heat of Ca^{2+} binding to DMPA bilayers in kcal mol^{-1} DMPA.

cations share the hydration shell with the negatively charged phosphates (Rialdi and Raffanti, 1984; Krakauer, 1972; Heinicke *et al.*, 1974). In the case of DMPA this positive contribution to ΔC_p is evidently overcompensated by another effect giving a decrease in C_p. We believe that the expulsion of water molecules from hydrophobic regions, i.e., a decrease in "hydrophobic hydration," is responsible for the overall negative ΔC_p observed for the binding of Ca^{2+} to DMPA. Ca^{2+} binding thus seems to induce a tighter packing of the acyl chains in the gel phase leading also to the observed increase in transition temperature. The sudden strong decrease in ΔH above 50°C corresponds to a situation where in addition to the binding of Ca^{2+} a phase transition from the liquid-crystalline L_α to the gel phase is induced. The measured heat of reaction thus includes the negative phase transition enthalpy ΔH.

Another negatively charged phospholipid is phosphatidylglycerol. As PG has a phosphodiester group, only one proton can be released from the phosphate. At neutral pH the PG head group is negatively charged, but it can be protonated by lowering the pH. The apparent pK is of course again dependent on the surface potential according to equation (31). With increasing degree of protonation the transition temperature increases (Watts *et al.*, 1978) while the transition enthalpy stays almost invariant. Binding of Ca^{2+} also leads to an increase in T_m but a slight decrease in ΔH is observed in this case. Figure 28 shows DSC scans of DPPG at pH 8.5 and 2.7 and with an equimolar concentration of Ca^{2+}. When the total reaction enthalpy for proton binding or Ca^{2+} binding is plotted as a function of temperature a

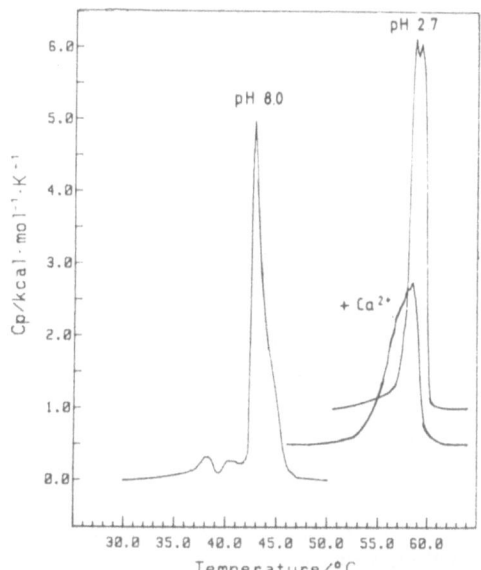

Figure 28. DSC scans of DPPG at pH 8.0 and 2.7 and with Ca^{2+} (1:1).

negative slope is obtained in both cases (see Fig. 29). Above 43°C the reaction enthalpy becomes suddenly strongly negative as proton or Ca^{2+} binding induces a transition from the L_α to an ordered phase. This is similar to the case of Ca^{2+} binding to DMPA discussed above. Below 40°C the ΔC_p values determined from the slopes are -35 cal K^{-1} mol^{-1} for proton binding and -45 cal K^{-1} mol^{-1} for Ca^{2+} binding. Again these values indicate that the effects of the liberation water from the polar head groups are overcompensated by the expulsion of water from hydrophobic regions when the lamellar structures become more ordered on proton or Ca^{2+} binding. Again we can compare the DPPG system with the protonation of glycero-1-phosphate where the effects from "hydrophobic hydration" are not operative. For this molecule the ΔC_p for protonation is positive with $+80$ cal K^{-1} mol^{-1}, a large difference compared to the -35 cal K^{-1} mol^{-1} measured for proton binding to DPPG. It could be that the change in tilt of the hydrocarbon chains induced by protonation (Watts *et al.*, 1978; 1981) may give an additional contribution.

The examples in this section show that reaction calorimetry can be successfully employed to measure heats of ion binding and to determine changes in heat capacity which reflect changes in structure, particularly changes in bilayer–water interactions. In addition titration experiments can be used to determine the modes of binding and the binding constants.

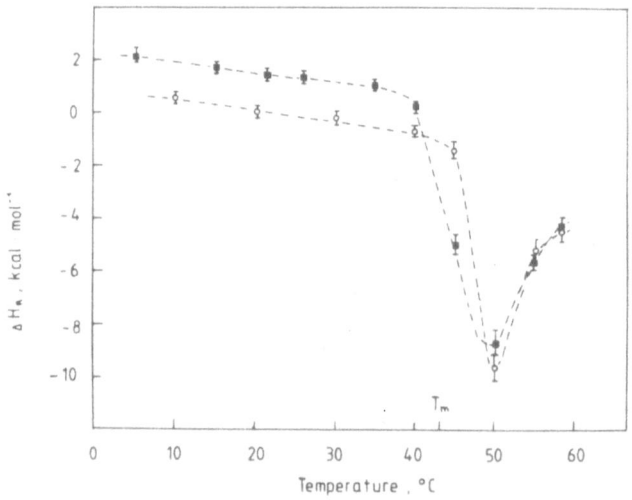

Figure 29. Temperature dependence of the heat of protonation (■) and Ca-binding (○) to DPPG in kcal mol^{-1} DPPG.

3.3. Determination of the Heat of Protein Incorporation into Bilayers

Another example of the application of reaction calorimetry is measurements of the heat of incorporation of model peptides into lipid bilayers. In Section 1.3.5, DSC scans of lipid–melittin mixtures were shown as a function of melittin concentration. It was mentioned that melittin is a water soluble polypeptide and is incorporated into the membrane when added to phospholipid vesicles. This offers the possibility of studying directly the incorporation reaction by titration calorimetry. The decrease of the transition enthalpy ΔH with increasing melittin content observed in the DSC scans (see Figs. 16 and 17) was assumed to be caused by the perturbation of the lipids surrounding the protein. The lipids in this annulus were supposedly in an energetically intermediate state compared to gel or liquid-crystalline lipids. In addition they were assumed to change their properties only continuously with temperature and not to take part in the transition. In most cases it cannot be determined from the DSC measurements whether this intermediate state is exactly in the middle of the two extremes or whether the annular lipid is closer to the liquid-crystalline lipid in its properties. In the first example of BR–DMPC mixtures it was assumed for the calculation of the transition curves that the annular lipid is more liquid-crystalline-like in its character (see Section 1.3.5).

The state in which these annular lipids exist can be expressed in the form of an order parameter S (Jähnig, 1981; Owicki and McConnell, 1979; Jähnig et al., 1982; Petersen and Chan, 1977). Figure 30 shows schematically the change of the order parameter S with temperature for free unperturbed lipid, where a large change in S is observed at the phase transition and two hypothetical curves for the order parameter of annular lipids, which do not undergo a transition but display only a continuous change in S. The upper curve corresponds to the case where the annular lipid is exactly in an intermediate state compared to free gel or liquid-crystalline lipid, and for the lower curve the annular lipid was assumed to be more liquid-crystalline-

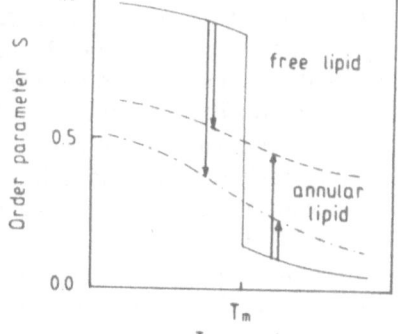

Figure 30. Schematic drawing of the change of the order parameter S for free lipids undergoing a phase transition at T_m (solid line) and for two hypothetical cases of annular lipid surrounding a protein not taking part in the phase transition.

like. In the first case the differences in order parameter just below and above T_m between free and annular lipid have the same magnitude, but they are different in the second case. This order parameter change is in some way related to changes in enthalpy. In an experiment where the enthalpy for protein incorporation could be measured just below and above T_m a distinction between these two hypothetical cases could be made, depending on whether or not the reaction enthalpies below and above T_m have the same magnitude. Jähnig (1983) has shown in a model calculation that the enthalpic contribution of an α-helix changing its surrounding from water to a hydrophobic bilayer is small compared to the enthalpy changes occurring in the lipids when they are transferred from free to annular lipid.

With melittin these experiments can be performed easily as this polypeptide is water soluble and incorporates readily into the bilayer, particularly above T_m in the liquid-crystalline state of the lipids. Figure 31 shows the integral heat of reaction, when DMPC is titrated with a melittin solution at temperatures above and below T_m. As expected from the model in Fig. 30, the sign of the reaction enthalpy changes below T_m. Above T_m negative heats of reaction are observed corresponding to a decrease in enthalpy (i.e., an increase in order), and below T_m the reaction enthalpy is positive, i.e., the order is decreased. So the experimental results support the model calculations of Jähnig (1983) and they agree with results reported for other lipid–protein systems (Massey *et al.*, 1981; Epand and Sturtevant, 1981).

We performed a similar set of experiments with the lipid DMPA where the DSC scans were shown in Fig. 17. Figure 32 shows the integral heats of reaction at two different temperatures for this system. The results are qualitatively the same as for the DMPC–melittin system in that the sign of the reaction enthalpy is different above and below T_m. It is evident, however, that the absolute values are different for the two systems and that the shapes of the curves are also different. This was to be expected, as the plots of ΔH versus melittin content (see Fig. 18) were not identical for the two systems, indicating different degrees of perturbation of annular lipids or melittin aggregation which depends on the nature of the phospholipid head group.

From the plots in Figs. 31 and 32 the incorporation enthalpy per mole of melittin can be calculated. The values obtained are from +30 to +60 kcal·mol^{-1} below T_m and from −20 to −40 kcal mol^{-1} above T_m. So the absolute values of the reaction enthalpy are smaller above than below T_m, indicating that the ordering effect of melittin on liquid-crystalline lipid is smaller than the perturbation effect on the gel phase lipid, a result that was also obtained for the BR–DMPC system. So these results obtained by reaction calorimetry also support an order parameter for annular lipids represented by the lower curve in Fig. 30. A note of caution should be added, however. The order parameter S in Fig. 30 is defined in a purely operational way and the relation

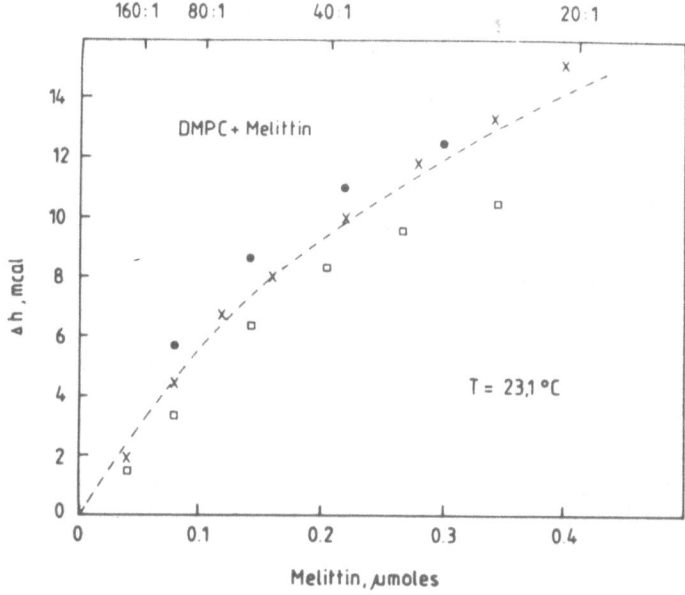

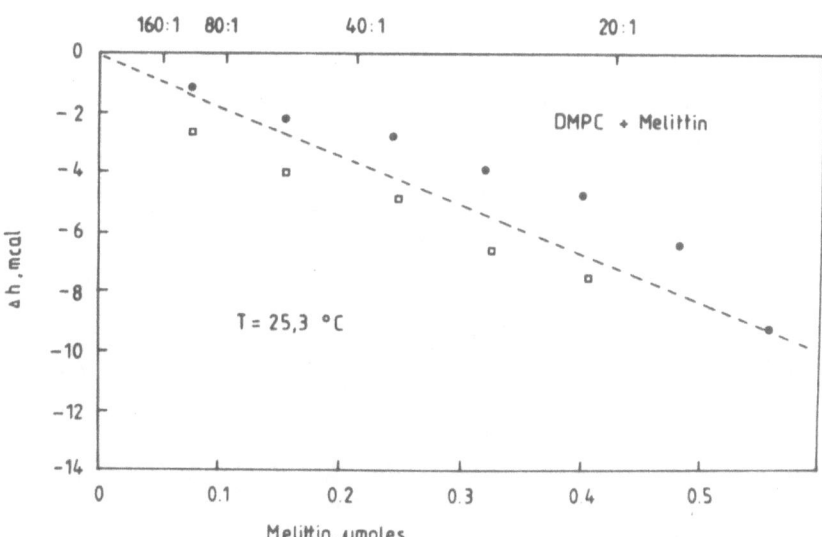

Figure 31. Integral heat of reaction for melittin incorporation into DMPC bilayers above and below the phase transition of DMPC. Different symbols represent separate experiments. Experimental conditions: 5 ml 1.6×10^{-3} M DMPC + 4×10^{-3} M melittin.

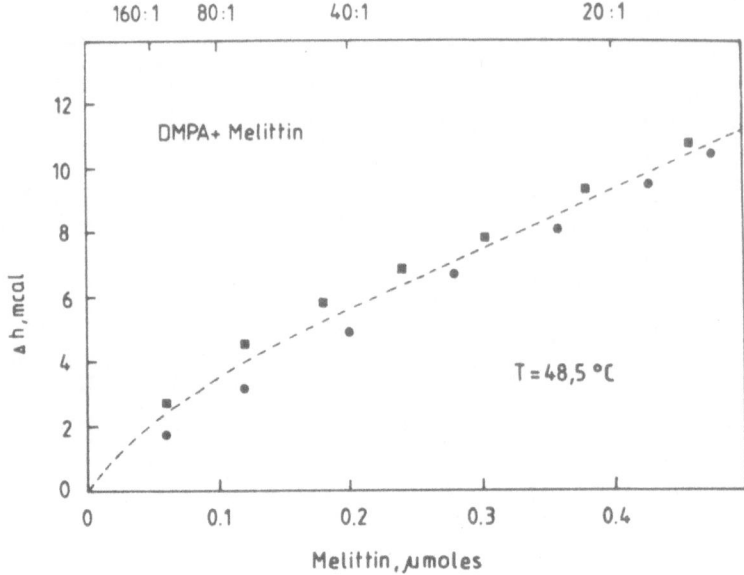

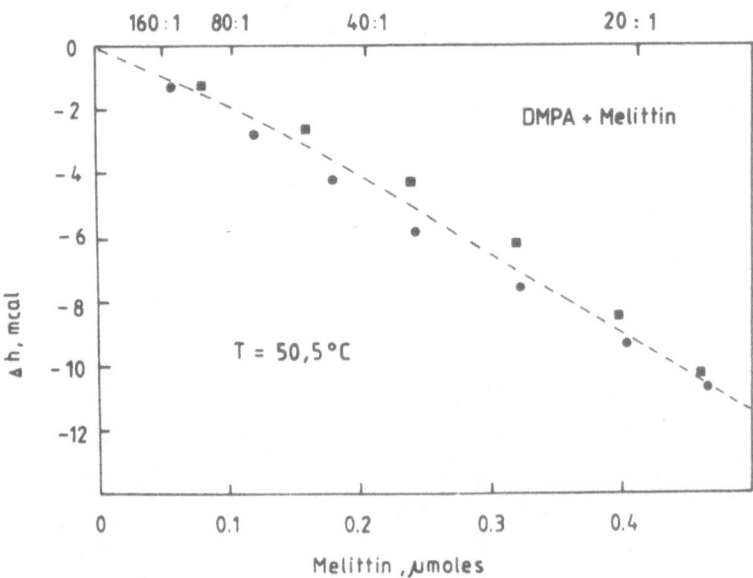

Figure 32. Integral heat of reaction for melittin incorporation into DMPA bilayers at pH 6 above and below the transition of DMPA. Different symbols represent separate experiments. Experimental conditions: 5 ml 1.6×10^{-3} M DMPA $+ 4 \times 10^{-3}$ M melittin.

to the enthalpy H of the system is not obvious. For a more detailed description of lipid behavior, S is usually set up as a product of the form $S = S_\alpha \cdot S_\beta$, where S_α is the conformational order parameter and S_β a rigid body order parameter describing the ordering of the director axis relative to the bilayer normal (Jähnig et al., 1982; Petersen and Chan, 1977; Jähnig, 1983; Meier et al., 1983). An increase in order parameter S can occur by changes in S_α and/or S_β. It is also possible that the conformational order may decrease but the rigid body order will increase giving an overall increase in S. To what extent changes in the enthalpy H reflect changes in S_α or S_β is not known. A decrease of the number of *gauche* conformations will of course result in a decrease in H, as will an increase in rigid body order, because van der Waals interactions are orientation dependent. Which of the two effects will dominate cannot be said a priori.

Below the phase transition measurements of melittin incorporation are possible only very close to the transition temperature as the kinetics of incorporation becomes very slow. Above T_m, however, a larger temperature interval is accessible. Figure 33 shows the reaction enthalpy as a function of temperature for DMPC and DMPA at a lipid/melittin ratio of 80:1. The reaction enthalpy decreases rapidly with increasing temperature. That this should occur was also suggested by the model calculations of Jähnig (1983).

Another prediction of the model, namely, that the changes in protein conformation or interaction do not contribute to a measurable extent to the reaction enthalpy, can also be tested. For this test the difference in transition enthalpy between pure lipid and the lipid-protein mixture at a certain lipid/protein ratio has to be determined from Fig. 18 and compared to the sum of the absolute values of the reaction enthalpies for the incorporation reaction measured just above and below T_m. It turns out that these two values are very similar, differing by not more than 5%-10%. The measurements thus support the model that the main contributions to the observed enthalpic effects are caused by changes in order of the phospholipids. In addition the concentration dependence of the enthalpy of incorporation

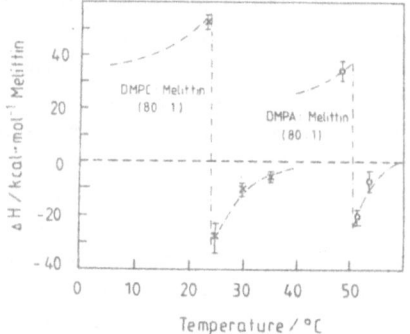

Figure 33. Temperature dependence of the heat of reaction for melittin incorporation into DMPC and DMPA bilayers at a fixed lipid:melittin ratio (80:1) in kcal mol⁻¹ melittin.

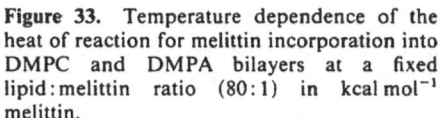

determined from Figs. 31 and 32 corresponds nicely to the changes in transition enthalpy shown in Fig. 18.

Titration experiments similar to the ones shown for DMPC and DMPA have been carried out with DPPG and DMPG. In this case the situation is more complicated as the transition temperature of the system increases owing to electrostatic effects when melittin is incorporated (see Fig. 19). In the titration experiments carried out just above T_m of the pure DMPG or DPPG this is manifested in a change of sign of the heat of reaction. First it is negative as observed before, but changes sign when the melittin concentration is increased because the transition temperature has increased and the temperature at which the experiment is performed is now below T_m. At lower temperatures, the reaction enthalpy is always positive with a value of ca. 25 kcal mol^{-1} melittin. In the case of PGs there are larger differences when the sums of the reaction enthalpies determined in the titration experiments are compared with the differences in the heats of transition between pure PG and PG–melittin mixtures. Apparently enthalpic contributions from changes in the peptide itself or changes in interactions are also present. In this case they may arise from electrostatic interactions between the negatively charged head groups and the protein. We have seen that in the PG–melittin mixtures these are responsible for the increase in transition temperature.

The examples of the application of reaction calorimetry presented in this section show that this method is very powerful and can give a very detailed picture of lipid–protein interactions, particularly in combination with data from DSC measurements on the same lipid–protein system. In addition the calorimetric binding curves contain information on the binding constants, which could be evaluated using appropriate models (Stankowski, 1983a,b).

4. CONCLUSIONS

Differential scanning calorimetry is a powerful tool to study the thermotropic behavior of single lipids and lipid mixtures. Together with theoretical calculations important information on lipid miscibility can be obtained. For studies of lipid–protein interactions DSC is a fast and easy-to-use technique to obtain a first characterization of the system before applying other more sophisticated techniques, which may need more material and may also be more time consuming.

Reaction calorimetry has so far not been used to a great extent in lipid work. The experimental results summarized in this chapter show that this calorimetric method is also very useful. Binding of ions to lipid bilayers can be studied in detail. The technique can also be applied to study the incorporation of amphiphilic peptides into bilayers obtaining additional information on lipid-protein interactions. The major drawback of reaction calorimetry is still its relatively low sensitivity. This comes mainly from the fact that the enthalpic effects in many reactions, which could in principle be followed by titration calorimetry, are very small, so that large amounts of material are needed. This will limit the application of this method to model membranes of synthetic lipids that are available in large quantities.

High-sensitivity DSC, however, is also applicable to biological membranes (8, 9), a topic beyond the scope of this chapter.

ACKNOWLEDGMENTS. Work from the author's laboratory was supported by grants from the Deutsche Forschungsgemeinschaft through grant No. SFB 60/D-4 and the Fonds der Chemischen Industrie and was done in collaboration with W. Hübner, P. End and M. Hillman. The DSC work on BR–DMPC was done in collaboration with M. Heyn and N. Dencher. Many of the experiments described in this chapter could not have been performed without the expert technical assistance of D. Hartenthaler and O. Thorwart, which is gratefully acknowledged.

REFERENCES

Bach, D., 1984, in: *Biomembrane Structure and Function* (Topics in Molecular and Structural Biology, D. Chapman, ed.), p. 1, Verlag Chemie, Weinheim.

Barisas, B. G., and Gill, S. J., 1978, *Ann. Rev. Phys. Chem.* 29:141.

Bernard, E., Faucon, J. F., and Dufourcq, J., 1982, *Biochim. Biophys. Acta.* 688:152.

Blazyk, J. F., Melchior, D. L., and Steim, J. M., 1975, *Anal. Biochem.* 68:586.

Blume, A., 1980, *Biochemistry* 19:4908.

Blume, A., 1983, *Biochemistry* 22:5436.

Blume, A., unpublished.

Blume, A., and Eibl, H., 1979, *Biochim. Biophys. Acta* 558:13.

Blume, A., and Griffin, R. G., 1982, *Biochemistry* 21:6230.

Blume, A., and Hillmann, M., 1986, *Eur. Biophys. J.* 13:343.

Calhoun, W. I., and Shipley, G. G., 1979, *Biochemistry* 18:1717.

Cevc, G., and Marsh, D., 1985, *Biophys. J.* 47:21.

Chapman, D., 1965, in: *The Structure of Lipids*, Methuen and Co., London.

Chapman, D., 1968, in *Biological Membranes* (D. Chapman, ed.), Vol. 1, p. 165, Academic Press, New York.

Chapman, D., Williams, R. M., and Ladbrooke, B. D., 1967, *Chem. Phys. Lipids* 1:445.

Chapman, D., Urbina, J., and Keough, K. M., 1974, *J. Biol. Chem.* 249:2512.

Chen, A., and Wadsö, I., 1982, *J. Biochem. Biophys. Methods* 6:307.

Chen, S. C., Sturteveant, J. M., and Gaffney, B. J., 1980, *Proc. Natl. Acad. Sci.* USA 77:5060.

Cohen, J. A., and Cohen, M., 1981, *Biophys. J.* 36:623.

Datta, S. P., and Grzybowski, A. K., 1958, *Biochem. J.* 69:218.

Demel, R. A. and de Kruyff, B., 1976, *Biochim. Biophys. Acta* 457:109.

Donovon, J. W., 1984 *Trends Biochem. Sci.* 1984:340.

von Dreele, P. H., 1978, *Biochemistry* 17:3939.

Eibl, H., 1984, *Angew. Chem.* 96:247.

Eibl, H., and Woolley, P., 1979, *Biophys. Chem.* 10:261.

Elamrani, K., and Blume, A., 1984, *Biochim. Biophys. Acta* 769:578.

Engel, J., and Schwarz, 1970, *Angew. Chem.* 82:468.

Epand, R. M., and Sturtevant, J. M., 1981, *Biochemistry* 20:4603.

Estep, T. N., Mountcastle, D. B., Biltonen, R. L., and Thompson, T. E., 1978, *Biochemistry* 17:1984.

Estep, T. N., Mountcastle, D. B., Barenholz, Y., Biltonen, R. L., and Thompson, T. E., 1979, *Biochemistry* 18: 2112.

Freire, E., and Snyder, B., 1980, *Biochemistry* 19:88.

Galla, H. J., and Sackmann, E., 1975, *J. Am. Chem. Soc.* 97:4114.

Gevod, V. S., and Birdi, K. S., 1984, *Biophys. J.* 45:1079.

Guggenheim, E. A., 1937, *Trans. Faraday Soc.* 33:151.

Guggenheim, E. A., 1944, *Proc. R. Soc. (London) Ser. A* 183:203, 213.

Habermann, E., 1972, *Science* 177:324.

Hauser, H., 1975, in: *Water: A Comprehensive Treatise* (F. Franks, ed.), Vol. 4, p. 209, Plenum Press, New York.

Hauser, H., Darke, A., and Phillips, M. C., 1976, *Eur. J. Biochem.* 62:335.

Heinicke, M., Bode, D., and Schernau, U., 1974, *Biopolymers* 13:227.

Heyn, M. P., Blume, A., Rohorek, M., and Dencher, N. A., 1981a, *Biochemistry* 20:7109.

Heyn, M. P., Cherry, R. J. and Dencher, N. A., 1981b, *Biochemistry* 20:840.

Hildebrand, H., 1929, *J. Am. Chem. Soc.* 51:66.

Hitchcock, P. B., Mason, R., Thomas, K. M., and Shipley, G. G., 1974, *Proc. Natl. Acad. Sci. USA* 71:3036.

Hui, S. W., and He, N.-B., 1983, *Biochemistry* 22:1159.

Ito, T., and Ohnishi, S., 1974, *Biochim. Biophys. Acta* 352:29.

Jähnig, F., 1981, *Biophys. J.* 36:329.

Jähnig, F., 1983, *Proc. Natl. Acad. Sci. USA* 80:3691.

Jähnig, F., Harlos, K., Vogel, H., and Eibl, H., 1979, *Biochemistry* 18:1459.

Jähnig, F., Vogel, H., and Best, L., 1982, *Biochemistry* 21:6790.

Krakauer, H., 1972, *Biopolymers* 11:811.

Kresheck, G. G., 1975, in: *Water: A Comprehensive Treatise* (F. Franks, ed.), Vol. 4, p. 95, Plenum Press, New York.

Krishan, K. S., and Brandts, J. F., 1978, *Methods Enzymol.* 49:1.

Ladbrooke, B. D., and Chapman, D., 1969, *Chem. Phys. Lipids* 3:304.

Ladbrooke, B. D., Williams, R. M., and Chapman, D., 1968, *Biochim. Biophys. Acta* 150:333.

Lee, A. G., 1977a, *Biochim. Biophys. Acta* 472:237.

Lee, A. G., 1977b, *Biochim. Biophys. Acta* 472:285.

Lee, A. G., 1978, *Biochim. Biophys. Acta* 507:433.

Lentz, B. R., Barrow, D. A. and Hoechli, M., 1980, *Biochemistry* 19:1943.

Mabrey, S., and Sturtevant, J. M., 1976, *Proc. Natl. Acad. Sci. USA* 73:3862.

Mabrey, S., and Sturtevant, J. M., 1978, in *Methods in Membrane Biology*, Vol. 9, p. 237, Plenum Press, New York.

Mabrey, S., Mateo, P. L., and Sturtevant, J. M., 1978, *Biochemistry* 17:2464.

McElhaney, R. N., 1982, *Chem. Phys. Lipids* 30:229.

McIntosh, T. J., 1980, *Biophys. J.* 29:237.

Marsh, D., and Watts, A., 1982, in *Lipid–Protein Interactions* (P. C. Jost and O. H. Griffith, eds.), Vol. II, J. Wiley, New York.

Mason, J. T., Broccoli, A. V., and Huang, C., 1981, *Anal. Biochem.* 113:96.

Massey, J. M., Gotto, A. M., and Pownall, H. J., 1981, *Biochemistry* 20:1575.

Meier, P., Ohmes, E., Kothe, G., Blume, A., Weidner, J., and Eibl, H., 1983, *J. Phys. Chem.* 87:4904.

Melchior, D. L., and Morowitz, H. J., 1972, *Biochemistry* 11:4558.
Menger, F. M., 1979, *Acc. Chem. Res.* 12:111.
Mollay, Ch., 1976, *FEBS Lett.* 64:65.
Nagle, J. F., and Wilkinson, D. A., 1978, *Biophys. J.* 23:159.
Nichols, N., Sköld, R., Spink, C., Suurkuusk, J., and Wadsö, I., 1976, *J. Chem. Thermodynam.* 8:1081.
Ohki, S., and Sauve, R., 1978, *Biochim. Biophys. Acta* 511:377.
Owicki, J. C., and McConnell, H. H. M., 1979, *Proc. Natl. Acad. Sci. USA* 76:4750.
Papahadiopoulos, D., 1974, *J. Colloid Interface Sci.* 58:459.
Petersen, N. O., and Chan, S. I., 1977, *Biochemistry* 16:2657.
Privalov, P., and Khechinashvili, N. N., 1974, *J. Mol. Biol.* 86:556.
Privalov, P., Plotnikov, V. V., and Filimonov, V. V. 1975, *J. Chem. Thermodynam.* 7:41.
Puskin, J. S., and Martin, T., 1979 *Biochim. Biophys. Acta* 552:53.
Rialdi, G., and Raffanti, S., 1984, in: *Thermochemistry and Its Applications to Chemical and Biochemical Systems* (M. A. V. Ribeiro da Silva, ed.), p. 511, D. Reidel, Dordrecht.
Roux, G., Perron, G., and Desnoyers, J. E., 1978, *Can. J. Chem.* 56:2808.
Seddon, J. M., Harlos, K., and Marsh, D., 1983, *J. Biol. Chem.* 258:3850.
Seelig, J., 1982, in: *Lipid–Protein Interactions* (P. C. Jost and O. H. Griffith, eds.), Vol. II, J. Wiley, New York.
Sheetz, M. P., and Chan, S. I., 1972, *Biochemistry* 11:4573.
Singer, N. J., and Nicholson, G. L., 1972, *Science* 175:720.
Stankowski, S., 1983a, *Biochim. Biophys. Acta.* 735:341.
Stankowski, S., 1983b, *Biochim. Biophys. Acta.* 735:352.
Stümpel, J., Niksch, A., and Eibl, H., 1981, *Biochemistry* 20:662.
Stümpel, J., Eibl, H., and Niksch, A., 1983, *Biochim. Biophys. Acta.* 727:246.
Sugar, I. P., and Monticelli, G., 1983, *Biophys. Chem.* 18:281.
Terwilliger, T. C., Weissman, L., and Eisenberg, D., 1982, *Biophys. J.* 37:353.
Tosteson, M. T., and Tosteson, D. C., 1981, *Biophys. J.* 36:109.
Träuble, H., 1971, *Naturwissenschaften* 58:277.
Träuble, H., 1976, in *Structure of Biological Membranes* (S. Abramhamson and I. Pasher, eds.), p. 509, Plenum Press, New York.
Träuble, H., and Eibl, H., 1974, *Proc. Natl. Acad. Sci. USA* 71:214.
Träuble, H., Teubner, M., Woolley, P., and Eibl, H., 1976, *Biophys. Chem.* 4:319.
Vogel, H., Jähnig, F., Hoffmann, V., and Stümpel, J., 1983, *Biochim. Biophys. Acta* 733:201.
Watts, A., Harlos, K., Maschke, W., and Marsh, D., 1978, *Biochim. Biophys. Acta* 510:63.
Watts, A., Harlos, K., and Marsh, D., 1981, *Biochim. Biophys. Acta* 645:91.
Wilkinson, D. A., and Nagle, J. F., 1981, *Biochemistry* 20:187.
Wilkinson, D. A., and Nagle, J. F., 1982, *Biochim. Biophys. Acta.* 688:107.
Wittebort, R. J., Schmidt, C. F. and Griffin, R. G., 1981, *Biochemistry* 20:4223.
Woolley, P., and Teubner, M., 1979, *Biophys. Chem.* 10:335.

Chapter 4

Molecular Mobility in Membranes

D. Marsh

1. INTRODUCTION

The dynamic properties of membrane molecules, namely, the lipids and the proteins, can be divided into two groups: (1) lateral diffusion in the plane of the membrane, and (2) rotational diffusion, principally about the membrane normal (Fig. 1). In general the rotational motions are considerably faster than the translational motions, indicating that the two, although clearly related, are not tightly coupled. Rotation about an axis in the plane of the membrane, in the case of integral proteins, and transverse diffusion or flip-flop of the polar lipid molecules across the bilayer (Fig. 1) is extremely slow, if it occurs at all. Typical time scales for the various motions are indicated in Table I.

The fundamental molecular process that gives rise to the mobility or fluidity of the membrane components is rotational isomerism in the lipid chains. This is driven by the entropy gain from rotational disorder, which necessitates looser packing and hence greater fluidity of the lipid environment. For synthetic lipids a cooperative transition from a highly ordered to a fluid phase takes place at a characteristic temperature at which the entropy gain from rotational disorder balances the loss in cohesive energy

D. MARSH • Max-Planck-Institut für biophysikalische Chemie, Abteilung Spektroskopie, D-3400 Göttingen, Federal Republic of Germany.

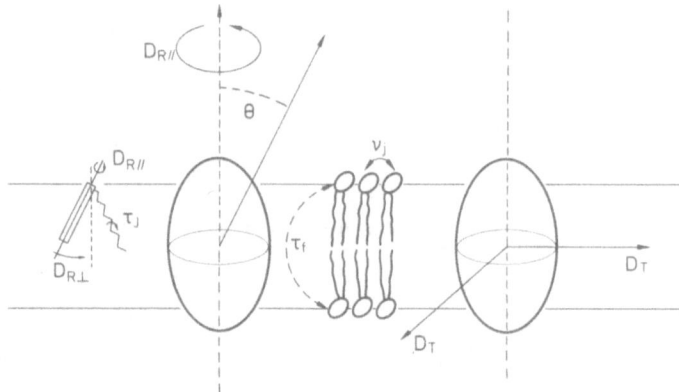

Figure 1. Motional modes of the lipid and protein molecules in a biological membrane. The protein molecules can rotate about an axis perpendicular to the plane of the membrane with rotational diffusion coefficient, $D_{R\parallel}$, and also possibly execute a limited wobbling motion with mean amplitude, θ. The protein molecules can also translate laterally within the plane of the membrane with translational diffusion coefficient, D_T. The lipid molecules can rotate about their long axis with rotational diffusion coefficient, $D_{R\parallel}$, and also perform a limited rotation about a perpendicular axis (within the plane of the membrane) with rotational diffusion coefficient, $D_{R\perp}$. The lipid chains can undergo *trans-gauche* isomerism about individual C–C bonds with a mean jump rate of τ_j^{-1}. The lipid molecules as a whole may also translate laterally with a mean jump frequency, ν_j. The trans-bilayer jump rate, τ_f^{-1}, for the lipid molecules is extremely slow.

from the looser chain packing. The chain isomerism rates are the fastest large-scale motions in the membrane (Table I).

2. LATERAL DIFFUSION

This section deals with the two-dimensional translational diffusion of protein and lipid molecules in the plane of the membrane. In fluid membranes the diffusion is likely to be isotropic within the plane of the membrane, at least to a first approximation. First the translational diffusion

Table I. Typical Time Scales for the Motion of Membrane Components in a Fluid Lipid Environment[a]

	D_T (cm^2 sec^{-1})	ν_j (sec^{-1})	$D_{R\parallel}$ (sec^{-1})	$D_{R\perp}$ (sec^{-1})	τ_j^{-1} (sec^{-1})	τ_f^{-1} (sec^{-1})
Protein	10^{-8}	—	10^5	~ 0	—	~ 0
Lipid	10^{-7}	10^8	10^9	10^9	10^{10}	10^{-5}

[a] D_T and ν_j refer to the translational diffusion of proteins and lipids, respectively. $D_{R\parallel}$ and $D_{R\perp}$ refer to rotational diffusion. τ_j^{-1} refers to the *trans-gauche* isomerization in the lipid chains, and τ_f^{-1} refers to the trans-membrane flip–flop of the lipid molecules (see Fig. 1).

coefficient is defined via the two-dimensional diffusion equation derived from Fick's law. Next the solutions of the diffusion equation are presented for two commonly used methods for determining the diffusion coefficient, namely, fluorescence photobleaching and the measurement of bimolecular collision rates. Finally two sets of models are presented for interpreting the translational diffusion coefficient. Of these, the hydrodynamic models are more appropriate to protein diffusion and the free volume models to diffusion of the lipid molecules.

2.1. Translational Diffusion Equation

The translational diffusion of lipid and protein molecules within the plane of the membrane can be described by a two-dimensional diffusion equation:

$$\frac{\partial P}{\partial t} = D_T \nabla^2 P$$

$$= D_T \left[\frac{\partial^2}{\partial r^2} + \frac{1}{r} \frac{\partial}{\partial r} \right] P \tag{1}$$

where $P(r, t)$ is the probability that a given molecule is situated at distance r from the origin at time t, and D_T is the lateral diffusion coefficient. Fluorescence photobleaching (FRAP) experiments have demonstrated that the diffusion in fluid lipid bilayers is radially isotropic, as indicated by equation (1). The solution of the diffusion equation for the initial boundary condition $P(r, 0) = \delta(r)$ is given by

$$P(r, t) = (4\pi D_T t)^{-1} \exp(-r^2/4D_T t) \tag{2}$$

The corresponding solution for the three-dimensional case is

$$P(r, t) = (4\pi D_T t)^{-3/2} \exp(-r^2/4D_T t) \tag{3}$$

When the origin of coordinates ($r = 0$) is not a fixed position, but is defined by the position of another identical molecule, D_T must be replaced by $2D_T$ in equations (2) and (3). This is the required solution when the diffusion coefficient is measured using a bimolecular interaction process such as excimer formation, luminescence quenching, or magnetic spin–spin interaction.

Either equation (2) or a two-dimensional random walk model may be used to determine the mean square displacement, $\langle r^2 \rangle$, of the diffusing particle in time t'. This is given by the Einstein relation:

$$\langle r^2 \rangle = 4D_T t' \tag{4}$$

Alternatively if a lattice model is used for the diffusion process, the diffusion coefficient is given by

$$D_T = \tfrac{1}{4}\nu_j\lambda^2 \tag{5}$$

where λ is the lattice constant (or mean free path) and ν_j is the hopping frequency between adjacent lattice sites. Again for bimolecular diffusive encounters D_T must be replaced by $2D_T$ in equations (4) and (5).

2.2. Fluorescence Photobleaching Recovery

In the fluorescence recovery after photobleaching (FRAP) method the fluorophores in a small area of the membrane are destroyed photochemically and then the rate of recovery of fluorescence in this area is monitored as a function of time, as unbleached fluorophores diffuse into the bleached region. A relatively small fraction of the total membrane molecules is fluorescently labeled. The method can be applied to lipids, proteins, or other membrane-associated molecules.

If a circular disk of radius w is uniformly bleached, the bleaching intensity profile is given by

$$I(r) = \begin{cases} I_0, & r \le w \\ 0, & r > w \end{cases} \tag{6}$$

where I_0 depends on the (uniform) laser beam intensity. The distribution of unbleached fluorophores at time $t = 0$ after bleaching is then given by

$$P(r, 0) = [1 - \alpha I(r)]P_0 \tag{7}$$

where P_0 is the initial, uniform fluorophore probability density (per unit area) before bleaching, and α is an expression depending on the duration and efficiency of bleaching, with a limiting value $\alpha I_0 = 1$ corresponding to complete bleaching of the circular region.

The fluorescence intensity observed at time t after bleaching is then given by

$$F(t) = \text{const} \times \int I(r)P(r, t)\, d^2r \tag{8}$$

where "const" is a constant that depends on the quantum efficiency of detection and the attenuation of the beam used for observation after bleaching. Here $I(r)$ is proportional to the observation intensity profile, which, to within the attenuation factor, is the same as that used for the initial bleaching. $P(r, t)$ is given by the solution of the two-dimensional diffusion equation [equation (1)], subject to the initial condition in equation (7) and the boundary condition:

$$P(\infty, t) = P_0 \tag{9}$$

which states that the fluorophore concentration at large distances is affected neither by the bleaching nor by the subsequent redistribution. The solution has been given by Axelrod et al. (1976) in terms of the fractional recovery in fluorescence intensity at time t:

$$f(t) = \frac{F(t) - F(0)}{F(\infty) - F(0)} \tag{10}$$

For bleaching and observation with a uniform circular profile the fractional recovery is given by

$$f(t) = 1 - (\tau/t)\exp(-2\tau/t)[I_0(2\tau/t) + I_2(2\tau/t)]$$

$$+ 2\sum_{k=0}^{\infty} \frac{(-1)^k(2k+2)!(k+1)!(\tau/t)^{k+2}}{(k!)^2[(k+2)!]^2} \tag{11}$$

where I_0 and I_2 are modified Bessel functions and the characteristic time constant is given by: $\tau = w^2/4D_T$.

The time dependence of the fluorescence recovery, according to equation (11), is given in Fig. 2. The time required for 50% recovery is given by $\tau_{1/2} = 0.88\tau$. This gives a simple method for determining the diffusion constant from the recovery curves:

$$D_T = 0.22(w^2/\tau_{1/2}) \tag{12}$$

Axelrod et al. (1976) have also given expressions appropriate to bleaching and observation with a Gaussian intensity profile. This is of practical importance since the laser beam intensity profile is commonly intermediate between Gaussian and a uniform circular disk, owing to admixture of the TEM 00 and TEM 01 laser modes. For a Gaussian profile equation (12) holds, with the factor of 0.88 replaced by a constant γ, which depends on the exact details of the bleaching intensity and its profile. The same authors have also investigated the effects of unidirectional lipid flow on the fluorescence recovery curves. In this case the diffusion equation [equation (1)] is replaced by

$$\frac{\partial P}{\partial t} = -V_0 \frac{\partial P}{\partial x} \tag{13}$$

where V_0 is the velocity of the uniform lipid flow in the x direction. As can be seen from Fig. 2, the two cases of translational diffusion and lipid flow may be distinguished from the differences in time course of the fluorescence recovery curves.

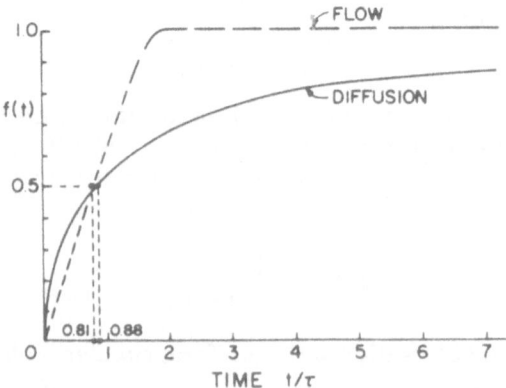

Figure 2. Fractional fluorescence recovery after photobleaching as a function of the reduced time, t/τ, for fluorophores initially randomly distributed in a planar membrane. Results are given for diffusion (——) and unidirectional flow (- - -), with bleaching and observation by a uniform circular beam (Azelrod *et al.*, 1976).

2.3. Bimolecular Collision Rates

Translational diffusion coefficients may also be determined from the bimolecular collision rates between labeled molecules. Methods used include excimer formation with pyrene-labeled lipids, spin–spin interactions between spin-labeled lipids, and luminescence quenching.

In principle the diffusion-controlled collision rate can be obtained from the solution of the diffusion equation [equation (1)]. A suitable boundary condition must be introduced at the position of the reference partner of the colliding pair, hence creating a concentration gradient. The Smoluchowski boundary condition assumes that the particle probability density goes to zero at the encounter surface:

$$P(d_c, t) = 0 \tag{14}$$

where d_c is the encounter distance. At large separations the boundary condition is that given previously in equation (9). With the initial condition of uniform probability density $P(r, 0) = P_0$, the solution to the two-dimensional diffusion equation is (Carslaw and Jaeger, 1959)

$$P(r, t) = \frac{2P_0}{\pi} \int_0^\infty \exp(-D_T u^2 t)$$

$$\times \frac{J_0(ud_c)\,Y_0(ur) - J_0(ur)\,Y_0(ud_c)}{J_0^2(ud_c) + Y_0^2(ud_c)}\,\frac{du}{u} \tag{15}$$

where J_0 and Y_0 are zero-order Bessel functions of the first and second kind, respectively. The collision frequency is given by the particle density flux across the perimeter of the circle of radius, d_c:

$$\tau_{coll}^{-1} = 4\pi D_T \left(r \frac{\partial P}{\partial r} \right)_{d_c}$$

$$= \frac{16 D_T P_0}{\pi} \int_0^\infty \exp(-D_T u^2 t) \frac{du}{u[J_0^2(ud_c) + Y_0^2(ud_c)]} \tag{16}$$

where τ_{coll} is the mean time between collisions. The time dependence of equation (16) is given in Fig. 3 in terms of the second-order collision rate constant: $k_{coll} = \tau_{coll}^{-1}/P_0$. At $t = 0$ the integral diverges, because of the particular form of the Smoluchowski boundary condition. For $t > 0$ the collision rate constant decreases steadily with time, without reaching a steady constant value. As pointed out by Razi Naqvi (1974) this is also true for the so-called "radiation" boundary condition, which assumes that the flux of the probability density at the encounter surface is given by

$$kP(d_c, t) = D_T \left(\frac{\partial P}{\partial r} \right)_{d_c} \tag{17}$$

where k has the nature of an effective second-order reaction rate constant. For this boundary condition the collision frequency does not diverge at $t = 0$, but nonetheless decreases steadily with time. Under these circumstances only an approximate value for the diffusion coefficient can be determined from the bimolecular collision rate. For instance, from Fig. 3 for Smoluchowski boundary conditions, the diffusion coefficient is given by $1-4 \times (k_{coll}/4\pi)$, for times in the range $1-10^3 \times d_c^2/D_T$.

The problem arises from the necessity of imposing a concentration gradient in order to apply Fick's law to the calculation of a collision rate in a system for which a steady-state concentration gradient need not exist. For lipid molecules a way out of this difficulty is provided by the quasicrystalline or hopping model of diffusion. If a given lipid exchanges at a frequency,

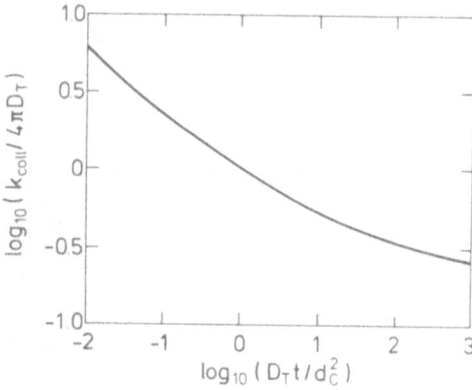

Figure 3. Bimolecular collision rate constant, k_{coll}, as a function of time, t, obtained from solution of the two-dimensional diffusion equation [equation (1)], with the Smoluchowski boundary condition for the particle probability density at the encounter distance (d_c): $P(d_c, t) = 0$.

ν_j, with one of its six nearest neighbors (Fig. 4), it encounters three new nearest neighbors and the collision frequency is given by (Devaux *et al.*, 1983)

$$\tau_{coll}^{-1} = 3\nu_j c \qquad (18)$$

where c is the mole fraction of labeled lipid molecules. Combining equation (18) with the Einstein expression equation (5) then gives the translational diffusion coefficient as

$$D_T = \left(\frac{\lambda^2}{24}\right)\left(\frac{\tau_{coll}^{-1}}{c}\right) \qquad (19)$$

where the factor of 2 appropriate to bimolecular encounters has been included in equation (5).

It is clear that the use of a hexagonal lattice is only an approximation for fluid lipid membranes, but the factor of 3 in equation (18) will come close to the mean number of new neighbors encountered per jump. A two-dimensional continuum approach (Träuble and Sackmann, 1972) leads to a quantitatively similar result. The collision frequency is then given by the product of the area swept out by the interaction diameter ($2d_c$) in unit time, times the density of labeled lipids, i.e.,

$$\tau_{coll}^{-1} = 2d_c l \frac{c}{A_L} \qquad (20)$$

where l is the integrated distance of travel in unit time, and A_L is the area per lipid molecule. Since $l = \nu_j \lambda$ for the hopping model, equation (20) then becomes

$$\tau_{coll}^{-1} = \left(\frac{2d_c \lambda}{A_L}\right) \nu_j c \qquad (21)$$

For typical values $\lambda = 8\ \text{Å}$, $A_L = 60\ \text{Å}^2$, and $d_c = 10\ \text{Å}$, the quantity given in the brackets in equation (21) has the value 2.7, which makes equation (20) numerically very similar to equation (18).

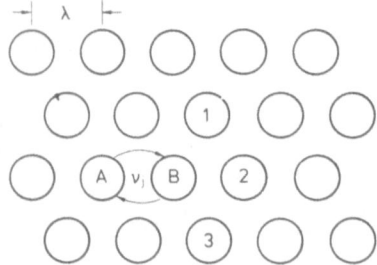

Figure 4. Quasihexagonal lattice model for the diffusion of lipid molecules in the plane of the membrane. Diffusion takes place by interchange of pairs of molecules, such as A and B, with frequency, ν_j. After interchange one molecule of the pair, A, has three new neighbors, 1-3. B may represent either a lipid molecule or a thermally induced void contributing to the total free volume.

2.4. Hydrodynamic Theories

The translational motion of protein molecules can be analyzed by the methods of classical hydrodynamics by assuming that the protein moves in a continuous two-dimensional viscous medium. The condition for the validity of this approach is that the diffusing species shall be much larger than the solvent (lipid) molecules. Problems inherent in the two-dimensional nature of the system can be alleviated by including consideration of the aqueous phase bounding the membrane (Clegg and Vaz, 1985).

Saffmann (1976) treated the system indicated in Fig. 5, in which a protein cylinder of radius a and height h is embedded in a membrane of uniform bulk viscosity, η_0 surrounded by aqueous phase of viscosity η'. Assuming that the viscosity of the aqueous phase is much smaller than that of the membrane, the frictional coefficient for the cylinder is given by

$$f_T = 4\pi\eta_0 h[-\gamma + \ln(\eta_0 h/\eta'a)]^{-1} \qquad (22)$$

where $\gamma = 0.5772$ is Euler's constant. It should be emphasized that the retardation of the protein diffusion in this model arises from the viscous drag of the external fluids on the membrane fluid (i.e., lipids) and not on the diffusing particle itself (Saffmann, 1976; Clegg and Vaz, 1985). In the absence of the external fluid, the mobility of the diffusing particle is infinite.

Hughes *et al.* (1981) have given the solution to the Saffmann model which is valid for any ratio of the aqueous and membrane viscosities. The asymptotic result, for the viscosity ratio $\varepsilon = (a/h)[(\eta_1' + \eta_2')/\eta_0] \gtrsim 1$, is given by

$$f_T = 4\pi\eta_0 h[\ln(2/\varepsilon) - \gamma + (4/\pi)\varepsilon - (\varepsilon^2/2)\ln(2/\varepsilon)]^{-1} \qquad (23)$$

where η_1' and η_2' are the viscosities of the two aqueous phases, and again $\gamma = 0.5772$ is Euler's constant. The exact result for arbitrary ε is given by Hughes *et al.* (1981). Saffmann's result is valid for $\varepsilon \gtrsim 0.1$.

Wiegel (1980) has extended the treatment to the case of a porous diffusing particle. This model might be applicable to the diffusion of large protein aggregates that allow some flow of lipid molecules within the

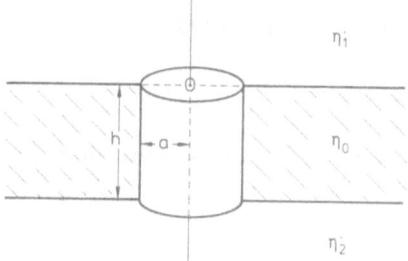

Figure 5. Hydrodynamic model for protein diffusion. A cylindrical protein molecule of radius, a, is embedded in a membrane of thickness, h, and uniform viscosity, η_0. The membrane is surrounded on its two sides by aqueous phases of viscosity η_1' and η_2', respectively.

aggregate. For a uniform permeable cylinder

$$f_T = 4\pi\eta_0 h\{-\gamma + \ln[\eta_0 h/\eta' a] + 2/\sigma^2 + I_0(\sigma)/\sigma I_1(\sigma)\}^{-1} \qquad (24)$$

where I_0 and I_1 are the zeroth- and first-order modified Bessel functions, and $\gamma = 0.5772$. Here $\sigma = a/k_0^{1/2}$ characterizes the permeability, where $k_0^{1/2}$ represents the depth to which the fluid flow penetrates. The additional term in equation (24) compared with equation (6) becomes $\pi\eta_0 h\sigma^2$ as $\sigma \to 0$, corresponding to the free draining limit, and goes to zero as $\sigma \to \infty$, corresponding to a totally impermeable protein.

Having obtained the frictional coefficients corresponding to either one of the three models given in equations (22)–(24), the translational diffusion coefficient is then given simply by the Einstein relation for Brownian diffusion:

$$D_T = kT/f_T \qquad (25)$$

2.5. Free Volume Theory

The free volume theory of diffusion in liquids (Cohen and Turnbull, 1959) can be applied to the diffusion of lipid molecules in membranes. The condition for the applicability of this model is that the diffusing molecules should be of comparable size or smaller than the solvent (in this case the lipid) molecules.

The mean free volume per lipid molecule is defined by

$$v_f = \langle v \rangle - v_0 \qquad (26)$$

where $\langle v \rangle$ is the mean volume occupied by one lipid molecule and v_0 is the van der Waals volume of the molecule. The probability that the lipid free volume rearranges to open up a void of volume v is given by statistical mechanical arguments as (Cohen and Turnbull, 1959)

$$p(v) = (\gamma/v_f) \exp(-\gamma v/v_f) \qquad (27)$$

where γ is a numerical factor to allow for possible overlap of free volumes ($\frac{1}{2} < \gamma < 1$). The total probability that the local free volume is greater than or equal to the critical value v^*, required for a diffusive displacement (Fig. 4) is then given by

$$P(v^*) = \int_{v^*}^{\infty} p(v)\, dv = \exp(-\gamma v^*/v_f) \qquad (28)$$

The translational diffusion coefficient is then

$$D_T = D(v^*) \exp(-\gamma v^*/v_f) \qquad (29)$$

where $D(v^*)$ is the diffusion coefficient in the void.

If the interfacial effects are neglected, $D(v^*)$ can be estimated from two-dimensional gas kinetic theory (Galla *et al.*, 1979):

$$D(v^*) = \tfrac{1}{4}\lambda^* \langle u \rangle \tag{30}$$

where λ^* is the distance of free travel in the volume v^*, and $\langle u \rangle$ is the mean velocity of the lipid molecule in the void. The mean distance of free travel is approximately equal to the intermolecular spacing: $\lambda^* \approx \lambda$, and the mean two-dimensional velocity is given by gas kinetic theory as $\langle u \rangle = (2kT/m)^{1/2}$, where m is the mass of the lipid molecule. An alternative formulation for the diffusion coefficient in the void, which takes into account the effective interfacial viscosities, has been given by Vaz *et al.* (1985). Here $D(v^*)$ is replaced by the Einstein relation

$$D(v^*) = kT/f_T^*$$

where f_T^* is the translational frictional coefficient in the void.

The free volume is dependent on both the temperature, T, and the externally applied pressure, ΔP. If $\bar{\alpha}_v$ and $\bar{\kappa}_v$ are the mean values of the thermal expansion coefficient and the bulk compressibility, respectively, over the temperature and pressure ranges of interest, then we have

$$v_f = \bar{\alpha}_v \langle v_m \rangle (T - T_0) - \bar{\kappa}_v \langle v_p \rangle \, \Delta P \tag{31}$$

where $\langle v_m \rangle$ and $\langle v_p \rangle$ are the mean values of the molecular volumes over the temperature range $T - T_0$, and the pressure range ΔP, respectively. T_0 is the temperature at which the free volume, in the absence of applied pressure, extrapolates to zero.

Finally, if the formation of the local free volume is an activated process, then this also contributes to the temperature dependence. An additional multiplicative term must be included in equation (29) and the diffusion coefficient becomes (Macedo and Litovitz, 1965)

$$D_T = D(v^*) \exp(-\gamma v^*/v_f - E_v^*/RT) \tag{32}$$

where E_v^* is the activation energy for the void formation.

3. ROTATIONAL DIFFUSION

This section deals with the rotational motion and the ordering of the lipid and protein molecules in the membrane. The rotational motion is specified by the rotational diffusion coefficients and the molecular order by the lipid order parameters. The amphiphilic structure of the membrane molecules dictates that the rotational motion is highly anisotropic and of limited amplitude with respect to the membrane normal (see Fig. 1). This somewhat complicates the solution of the diffusion equation. The section begins with discussion of the solution of the various forms of the rotational diffusion equation. Purely for illustrative purposes, isotropic rotation is included as a simple (although unrealistic) example. Then the effects of molecular ordering on the rotational motion are introduced. This is followed by an analysis of two commonly used methods for determining the rotational motion of membrane molecules, namely, luminescence depolarization and magnetic resonance spectroscopy. Both methods involve the determination of rotational correlation functions, which are obtained from solution of the rotational diffusion equation. Finally hydrodynamic models for the interpretation of protein rotational diffusion coefficients are discussed.

3.1. Rotational Diffusion Equation

Isotropic rotational diffusion can be described by the random diffusion of a point, corresponding to the end of a vector fixed in the molecule, over the surface of a unit sphere (see Fig. 6). The rotational diffusion equation expressed in the spherical polar angles (θ, ϕ) is then

$$\frac{\partial p}{\partial t} = D_R \nabla^2_\Omega p$$

$$= D_R \left[\frac{1}{\sin \theta} \frac{\partial}{\partial \theta} \left(\sin \theta \frac{\partial}{\partial \theta} \right) + \frac{1}{\sin^2 \theta} \frac{\partial^2}{\partial \phi^2} \right] p \qquad (33)$$

where $p(\Omega, t)$ is the instantaneous orientational distribution with respect to the polar angles $\Omega \equiv (\theta, \phi)$, and D_R is the isotropic rotational diffusion coefficient. The solution to equation (33) can be expressed in terms of a sum of spherical harmonics, $Y_L^M(\theta, \phi) = [(2L + 1)/4\pi]^{1/2} P_L(\cos \theta) e^{iM\phi}$, where $P_L(\cos \theta)$ is a Legendre polynomial:

$$p(\Omega, t) = \sum_{L,M} c_{LM}(t) Y_L^M(\Omega) \qquad (34)$$

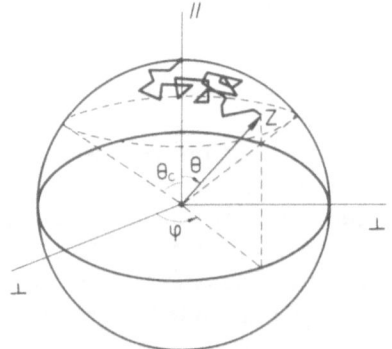

Figure 6. Rotational diffusion of a lipid or protein molecule. In the case of isotropic motion the locus of the principal molecular axis, Z, performs a random walk over the surface of a sphere. For anisotropic rotation, as in membranes, rotations about the Z axis, i.e., the third Euler angle, ψ, must also be considered. In addition the molecular order relative to the membrane normal ($\parallel$), restricts the angular amplitude, θ, by means of an orientational pseudopotential, $U(\theta)$. This is indicated schematically by a cone of maximum angle, θ_c.

The coefficients of the expansion $c_{LM}(t)$, are time dependent and can be determined by the initial condition: $p(\Omega, t) = \delta(\Omega - \Omega_0)$. The conditional probability that a molecule with orientation Ω_0 at time $t = 0$ subsequently has orientation Ω at time t is then given by

$$p(\Omega_0; \Omega, t) = \sum_{L,M} Y_L^{M^*}(\Omega_0) Y_L^M(\Omega) e^{-t/\tau_L} \tag{35}$$

where the time constant τ_L, is given by

$$1/\tau_L = L(L+1)D_R \tag{36}$$

Equation (35) will be required subsequently for the calculation of the autocorrelation functions, and we shall see that the τ_L's are in fact correlation times.

For anisotropic rotation, which is the situation more appropriate to membrane systems, the diffusion equation is

$$\frac{\partial p}{\partial t} = -\mathbf{L} \cdot \mathbf{D}_R \cdot \mathbf{L} p \tag{37}$$

where $\mathbf{L}$ is the operator which generates an infinitesimal elementary rotation (formally equivalent to the dimensionless angular momentum operator) and $\mathbf{D}_R$ is the rotational diffusion tensor. For axially symmetric molecular rotation, the diffusion equation can be written explicitly in terms of the Euler angles: $\Omega \equiv (\phi, \theta, \psi)$:

$$\frac{\partial p}{\partial t} = D_{R\perp}\left\{ \frac{1}{\sin\theta}\frac{\partial}{\partial\theta}\left(\sin\theta\frac{\partial}{\partial\theta}\right) + \frac{1}{\sin^2\theta}\frac{\partial^2}{\partial\phi^2} - \frac{2\cos\theta}{\sin^2\theta}\frac{\partial^2}{\partial\phi\,\partial\psi} \right.$$
$$\left. + \left(\cot^2\theta + \frac{D_{R\parallel}}{D_{R\perp}}\right)\frac{\partial^2}{\partial\psi^2}\right\} p \tag{38}$$

where $D_{R\parallel}$ and $D_{R\perp}$ are the rotational diffusion coefficients about the molecular symmetry axis and perpendicular to it, respectively (Fig. 1). The solution of equation (38) can be expressed as a linear combination of the normalized Wigner rotation matrices or generalized spherical harmonics: $\psi_{K,L}^M(\Omega) = [(2L+1)/8\pi^2]^{1/2}\mathscr{D}_{K,M}^L(\Omega)$, where $\mathscr{D}_{K,M}^L(\Omega)$ are the Wigner rotation matrices. Formally these are the eigenfunctions for a symmetric top, and for $K = 0$, they are related to the spherical harmonics by $\mathscr{D}_{0,M}^L(\phi, \theta, \psi) = [4\pi/(2L+1)]^{1/2} Y_L^M(\theta, \psi)$.

The conditional probability can be expressed in a similar manner as for the isotropic case:

$$p(\Omega_0; \Omega, t) = \sum_{K,L,M} \psi_{K,L}^{M^*}(\Omega_0)\psi_{K,L}^M(\Omega) e^{-t/\tau_{K,L}} \tag{39}$$

where the correlation times are now given by

$$1/\tau_{L,K} = D_{R\perp}L(L+1) + (D_{R\parallel} - D_{R\perp})K^2 \tag{40}$$

3.2. Orientational Order

In the case of membrane-bound molecules, the rotation about the perpendicular axis $(D_{R\perp})$ is usually restricted by an orientational pseudopotential, $U(\Omega)$, which gives rise to molecular ordering relative to the membrane normal. In this case an additional term must be added to the diffusion equation, which then becomes (Favro, 1965)

$$\frac{\partial p}{\partial t} = -[\mathbf{L} \cdot \mathbf{D}_R \cdot \mathbf{L} + \mathbf{L} \cdot \mathbf{D}_R \cdot \mathbf{L}U(\Omega)/kT]p \tag{41}$$

If the orientational potential depends only on the angle, θ, between the molecular principal axis and the membrane normal, then equation (41) has the form

$$\frac{\partial p}{\partial t} = -\left[\mathbf{L} \cdot \mathbf{D}_R \cdot \mathbf{L} + \frac{1}{kT}\frac{D_\perp}{\sin\theta}\frac{\partial}{\partial\theta}\sin\theta\frac{\partial U(\theta)}{\partial\theta}\right]p \tag{42}$$

where the first term on the right-hand side is given by equation (38).

In general the orientational pseudopotential can be expanded in terms of the Wigner rotation matrices, $\mathscr{D}^L_{K,M}(\phi, \theta, \psi)$. For axially symmetric systems, the terms up to second order are (Freed, 1976):

$$U(\Omega) = \varepsilon_0\mathscr{D}^2_{0,0}(\Omega) + \varepsilon_2[\mathscr{D}^2_{2,0}(\Omega) + \mathscr{D}^2_{-2,0}(\Omega)] \tag{43}$$

where the ε_i represent the strength of the orientating potential. Explicitly in terms of the Euler angles, (ϕ, θ), this is given by

$$U(\phi, \theta) = \lambda\cos^2\theta + \varepsilon\sin^2\theta\cos 2\phi \tag{44}$$

where the first term is the leading term and corresponds to the usual Maier–Saupe (1959) pseudopotential, and the second term accounts for the nonaxiality in the alignment of the x and y molecular axes. The strengths of the pseudopotential terms, λ and ε, are of course directly related to the coefficients ε_i in equation (43).

The degree of ordering is specified by the order parameters, which are expressed in terms of time-averages of the direction cosines of the molecular axes with respect to the membrane normal. For the Cartesian coordinate system indicated in Fig. 7, the elements of the ordering tensor are defined by

$$S_{ij} = \tfrac{1}{2}(3\langle\cos\theta_i\cos\theta_j\rangle - \delta_{ij}) \tag{45}$$

where $i, j = x, y, z$ and the angle brackets indicate a time average over the rotational motion. The orthogonality relation for the direction cosines requires that the three principal values of the ordering tensor are related by $\sum_i S_{ii} = 0$. The elements of the ordering tensor can also be expressed as linear combinations of time-average values of the Wigner rotation matrices.

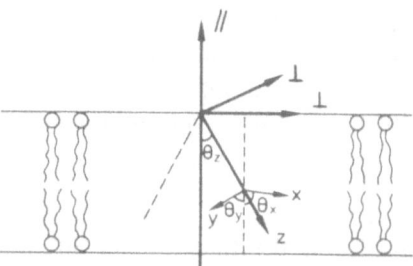

Figure 7. Instantaneous orientation of the molecular axes (x, y, z) relative to the symmetry axis, ($\parallel$), of the molecular ordering, i.e., the membrane normal. The molecular rotational diffusion is such as to produce a time-average axial symmetry relative to the ordering axis.

For instance, the principal order parameter is given by $S_{zz} = \langle \mathcal{D}^2_{0,0} \rangle = \langle P_2 \rangle$. The mean, or time-average, values are determined directly by the orientational distribution function. This means that they can be obtained from the orientational pseudopotential, $U(\Omega)$. For example, the S_{zz} element is given by

$$S_{zz} = \langle \mathcal{D}^2_{0,0} \rangle = \frac{\int \mathcal{D}^2_{0,0}(\Omega) \exp[-U(\Omega)/kT] \, d\Omega}{\int \exp[-U(\Omega)/kT] \, d\Omega} \tag{46}$$

For the simple model of rotational diffusion restricted within a cone of fixed angle θ_c (see Fig. 6) the order tensor can be obtained analytically: $S_{zz} = \frac{1}{2} \cos \theta_c (1 + \cos \theta_c)$, and from the axial symmetry of the system $S_{xx} = S_{yy} = -\frac{1}{2} S_{zz}$.

3.3. Fluorescence and Phosphorescence Depolarization

Rotational diffusion rates can be determined from the decay in polarization of luminescence of a labeled molecule following excitation by plane polarized light. Fluorescent labels that have lifetimes shorter than $\sim 1 \, \mu\text{sec}$ may be used for studying lipid rotational motion, and phosphorescent labels with lifetimes in the millisecond time range can be used for studying protein rotation. In addition, photobleaching or flash photolysis with polarized light may be used to study rotational motion, from the decay of the resulting photodichroism.

The anisotropy in the fluorescence (phosphorescence) polarization is defined by

$$r(t) = \frac{I_{\parallel}(t) - I_{\perp}(t)}{I_{\parallel}(t) + 2I_{\perp}(t)} \tag{47}$$

where $I_{\parallel}(t)$ and $I_{\perp}(t)$ are the fluorescence intensities with polarization parallel and perpendicular, respectively, to the plane of polarization of the exciting light. In general the probability of excitation by plane polarized light is proportional to $|\mu_a \cdot e_i|^2$, where μ_a is the absorption dipole moment and e_i is a unit vector in the direction of polarization of the incident light.

The probability of emission of light whose plane of polarization lies along the direction of the unit vector, e_f, is proportional to $|\boldsymbol{\mu}_e \cdot e_f|^2$, where $\boldsymbol{\mu}_e$ is the emission dipole moment. The net fluorescent intensity for excitation along e_i and observation along e_f is therefore

$$I_{ef}(t) = \langle |\boldsymbol{\mu}_a(\Omega_0, 0) \cdot e_i|^2 |\boldsymbol{\mu}_e(\Omega, t) \cdot e_f|^2 \rangle F(t) \tag{48}$$

where $F(t)$ is the fluorescence decay, whose time dependence may be ignored if the fluorescence lifetime is much longer than the rotational relaxation time. The angle brackets indicate an average over all initial molecular orientations, Ω_0, and all final orientations, Ω, i.e.,

$$I_{ef}(t) = F(t) \int |\boldsymbol{\mu}_a(\Omega_0) \cdot e_i|^2 |\boldsymbol{\mu}_e(\Omega) \cdot e_f|^2 p(\Omega_0) p(\Omega_0; \Omega, t) \, d\Omega_0 \, d\Omega \tag{49}$$

where $p(\Omega_0)$ is the probability that the molecule originally has orientation Ω_0. The time dependence is thus contained in the conditional probabilities $p(\Omega_0; \Omega, t)$, which are solutions of the diffusion equation and were introduced in Section 3.1.

For isotropic rotation equation (35) must be used for the conditional probability. From the orthogonality properties of spherical harmonics only the $L = 2$ terms in the expansion give nonzero values of the correlation function, and the polarization anisotropy is

$$r(t) = \tfrac{2}{5} P_2(\cos \delta) e^{-6D_R t} \tag{50}$$

where δ is the angle between the absorption and emission dipoles. Thus the polarization has an initial value $r(0) = \tfrac{2}{5} P_2(\cos \delta)$ and decays exponentially to zero with a rotational correlation time $\tau_R = \tau_2 = 1/(6D_R)$, corresponding to the $L = 2$ correlation function.

For anisotropic rotation equation (39) must be used for the conditional probability. In this case the terms with $L = 2$ and $K = 0, 1, 2$ contribute to the correlation function, and the polarization anisotropy is given by (Ehrenberg and Rigler, 1972)

$$r(t) = \tfrac{2}{5} P_2(\cos \theta_a) P_2(\cos \theta_e) e^{-6D_{R\perp} t}$$
$$+ \tfrac{3}{10} \sin 2\theta_a \sin 2\theta_e \cos(\phi_a - \phi_e) e^{-D_{R\parallel} t - 5D_{R\perp} t}$$
$$+ \tfrac{3}{10} \sin^2 \theta_a \sin^2 \theta_e \cos 2(\phi_a - \phi_e) e^{-4D_{R\parallel} t - 2D_{R\perp} t} \tag{51}$$

where (θ_a, ϕ_a) and (θ_e, ϕ_e) are the polar angles of the absorption and emission dipoles in the molecular coordinate system. Clearly in this case several of the correlation times in equation (40) enter into the polarization anisotropy and the decay is multiexponential. Even in the case of uniaxial rotation ($D_{R\perp} = 0$), which is likely to be a good approximation for integral membrane proteins, the decay is biexponential with time constants of $(4D_{R\parallel})^{-1}$ and $(D_{R\parallel})^{-1}$. Thus for anisotropic rotation, a unique rotational correlation time cannot be defined and one must work with the individual components of the diffusion tensor. The initial polarization must of course be as in the isotropic case: $r(0) = \frac{2}{5}P_2(\cos \delta)$. For polarized flash photolysis or photobleaching experiments bleaching and observation are at the same wavelengths corresponding to the same transition moment, which leads to a slight simplification in equations (50) and (51).

For fluorescently labeled lipid molecules the restricted rotation within an orientational potential must also be taken into consideration. An important feature of this restricted motion is that the anisotropy does not decay to zero, as in the isotropic case, but to a limiting value $r(\infty)$, at long times. Solutions of the Smoluchowski equation [equation (41)] are required in calculating the appropriate autocorrelation functions of the Wigner rotation matrices. The general result has been given by Zannoni (1981):

$$r(t) = \tfrac{2}{5}P_2(\cos \theta_a)P_2(\cos \theta_e)\sum_K G_{K,0}(t)$$
$$+ \tfrac{3}{10}\sin 2\theta_a \sin 2\theta_e \cos(\phi_a - \phi_e)\sum_K G_{K,1}(t)$$
$$+ \tfrac{3}{10}\sin^2 \theta_a \sin^2 \theta_e \cos 2(\phi_a - \phi_e)\sum_K G_{K,2}(t) \tag{52}$$

where the correlation functions of the Wigner rotation matrices are

$$G_{K,M}(t) = \langle \mathscr{D}^{2*}_{K,M}(\Omega_0)\mathscr{D}^{2}_{K,M}(\Omega)\rangle \tag{53}$$

and the summations range over $K = 0, 1, 2$. The evaluation of the correlation functions requires a knowledge of the form of the orientational pseudopotential, $U(\Omega)$. An analytical solution in closed form cannot be given even for the simple diffusion-in-a-cone model (Kinosita et al., 1977). However, an exact solution can be given for the limiting anisotropy, observed at long times $r(\infty)$, which corresponds also to the static polarization observed for probes with long fluorescence lifetimes (Lipari and Szabo, 1980):

$$r(\infty) = \tfrac{2}{5}P_2(\cos \theta_a)P_2(\cos \theta_e)\langle P_2(\cos \theta)\rangle^2 \tag{54}$$

where $\langle P_2(\cos \theta)\rangle$ is simply the order parameter, S_{zz}. An approximate result for the time dependence of the anisotropy decay in the presence of an orienting potential is (Lipari and Szabo, 1980)

$$r(t) = \tfrac{2}{5}P_2(\cos \theta_a)P_2(\cos \theta_e)[\langle P_2\rangle^2 + (1 - \langle P_2\rangle^2)e^{-6D_{R\perp}t}]$$
$$+ \tfrac{3}{10}\sin 2\theta_a \sin 2\theta_e \cos(\phi_a - \phi_e)$$
$$\times [\langle P_2\rangle^2 + (1 - \langle P_2\rangle^2)e^{-5D_{R\perp}t}]e^{-D_{R\parallel}t}$$
$$+ \tfrac{3}{10}\sin^2 \theta_a \sin^2 \theta_e \cos 2(\phi_a - \phi_e)$$
$$\times [\langle P_2\rangle^2 + (1 - \langle P_2\rangle^2)e^{-2D_{R\perp}t}]e^{-4D_{R\parallel}t} \tag{55}$$

where $\langle P_2(\cos\theta)\rangle$ characterizes the molecular order. This expression contains five independent exponential decays and therefore it is unlikely that it can be fitted in its entirety. An alternative approximation involving only three exponential decays has been given by van der Meer *et al.* (1984).

3.4. Magnetic Resonance Spectra

Information regarding molecular rotation can be obtained from the motionally induced modulation of the angular anisotropy of the interactions in magnetic resonance spectra. In electron spin resonance (ESR) spectroscopy the anisotropy of the g values and hyperfine interactions of nitroxide spin labels is normally used. In nuclear magnetic resonance (NMR) the anisotropy of the ^{31}P or ^{13}C chemical shift, of the deuterium quadrupole splitting, or of the C^{13}–H dipolar coupling is most frequently used. The time scale of the motional sensitivity depends on the frequency equivalent of the anisotropic interactions being modulated. For ESR this lies in the nanosecond range and for NMR in the millisecond to hundred milliseconds range. The magnetic resonance methods are therefore normally restricted to the measurement of lipid rotational motions and protein segmental flexibility. Saturation transfer ESR, however, depends on the spin lattice relaxation time of the spin label, which lies in the microsecond time range. Hence this latter method is applicable to the study of protein rotational motion.

For each of the anisotropic interactions, the contribution to the spin Hamiltonian can be written in terms of spherical tensor operators which have similar transformation properties to the Wigner rotation matrices:

$$\mathcal{H}_1(t) = \sum_{L,M} F'_{L,-M} A'_{L,M} \tag{56}$$

where $F'_{L,-M}$ are the tensor components representing the strength of the interaction, and $A'_{L,M}$ are spin operators. The interactions normally of interest in magnetic resonance are second-rank tensors, which means that only terms up to $L = 2$ are included in equation (56). The symmetry of the tensors can also further restrict the terms that have to be considered. The primes indicate that the tensors in equation (56) are expressed in molecule-fixed axes. The $F'_{L,-M}$ are represented naturally in this molecular coordinate system, but the spin operators are quantized in a space-fixed coordinate system specified by the spectrometer magnetic field. The $A'_{L,M}$ must therefore be expressed in the space-fixed system via rotation matrices, $\mathcal{D}^L_{K,M}(\Omega')$, transforming the space-fixed system to the ordering axis system (usually the membrane normal) and further via rotation matrices, $\mathcal{D}^L_{K,M}(\Omega)$, transforming the ordering axis system to the molecular axis system:

$$A'_{L,M} = \sum_K \mathcal{D}^L_{K,M}(\Omega) \sum_{K'} \mathcal{D}^L_{K',K}(\Omega') A_{L,K'} \tag{57}$$

where $A_{L,K'}$ are the spin operators in the space-fixed system. The time dependence is then, as previously, contained solely in the rotation matrices, $\mathscr{D}^L_{K,M}(\Omega)$, which is given by the solution of the rotational diffusion equation, equation (38) or (41).

In the motional narrowing regime, the angular anisotropy of the magnetic interactions is partially averaged and the time dependence of the spin Hamiltonian appears as a perturbation:

$$\mathscr{H}_1(t) = \langle \mathscr{H}_1 \rangle + \{\mathscr{H}_1(t) - \langle \mathscr{H}_1 \rangle\} \tag{58}$$

where $\langle \mathscr{H}_1 \rangle$ represents the time-averaged component of the anisotropic interaction and $\mathscr{H}_1(t) - \langle \mathscr{H}_1 \rangle$ is the perturbation. The condition for this is that the molecular motions shall be fast relative to the frequency equivalent of the anisotropies that they are modulating. The time-average interaction can be represented in terms of the molecular order parameters given in equations (45) and (46). For spin label ESR the order parameter of the nitroxide z axis S_{zz}, can be expressed in terms of the experimentally observed ^{14}N hyperfine splittings, $A_\parallel$ and $A_\perp$, appropriate to the spectrometer magnetic field oriented parallel and perpendicular to the ordering axis, respectively (see, e.g., Marsh, 1985):

$$S_{zz} = \frac{A_\parallel - A_\perp}{A_{zz} - \frac{1}{2}(A_{xx} + A_{yy})} \tag{59}$$

where A_{xx}, A_{yy}, A_{zz} are the principal elements of the hyperfine tensor appropriate to rigid order as in a single crystal. For ^{2}H NMR the order parameter of the C–D bond, S_{CD}, can be expressed in terms of the experimentally observed quadrupole splitting. For the magnetic field oriented perpendicular to the ordering axis (which gives the major peaks in the spectrum of an unoriented sample), the quadrupole splitting is given by (Seelig, 1977)

$$\Delta\nu_{Q\perp} = \tfrac{3}{4}(e^2qQ/h)S_{CD} \tag{60}$$

where (e^2qQ/h) is the static quadrupole coupling constant. In both equation (59) and equation (60) it may be necessary to transform from the nitroxide or C–D bond system to the principal axis system of the molecular rotational diffusion tensor, if these are not colinear. To obtain the order parameter of the principal molecular axis this may require assumptions regarding the axial symmetry of the rotational diffusion system. For ^{31}P NMR spectroscopy, the chemical shift anisotropy, unlike the ^{2}H quadrupole interaction or the nitroxide hyperfine interaction, is nonaxial. The chemical shift anisotropy then must be expressed in terms of two independent order parameters (Seelig, 1978):

$$\Delta\sigma = \sigma_\parallel - \sigma_\perp = S_{xx}(\sigma_{xx} - \sigma_{yy}) + S_{zz}(\sigma_{zz} - \sigma_{yy}) \tag{61}$$

where σ_{xx}, σ_{yy}, σ_{zz} are the principal elements of the chemical shift anisotropy tensor and S_{xx}, S_{zz} are the order parameters of the corresponding principal axes. A similar relation holds for the g-value anisotropy in spin label ESR spectra (see, e.g., Marsh, 1985), but in this latter case all principal values of the ordering tensor may be determined, since S_{zz} is obtained directly from the hyperfine anisotropy as indicated in equation (59).

In the fast motional (or motional narrowing) regime the time-dependent perturbations determine only the magnetic relaxation times, T_1 and T_2, and not the resonance line positions. T_1 is the spin–lattice relaxation time which represents the rate at which the spin populations of the magnetic energy levels (i.e., the longitudinal magnetization) return to their equilibrium values after transitions are induced between them. T_2 is the spin–spin relaxation time which represents the rate at which the transverse magnetization dephases after it is induced by a transition. The spin–spin relaxation time determines the linewidths in the magnetic resonance spectrum. Both relaxation processes are induced by modulation of the angular anisotropy of the magnetic interactions by the molecular rotation, although in different ways. The modulation of the magnetic anisotropies is expressed in terms of the autocorrelation function:

$$G(\tau) = \langle \mathcal{H}_1^*(t)\mathcal{H}_1(t+\tau)\rangle \tag{62}$$

The frequency dependence of the relaxation times is then given directly in terms of the spectral densities, $J(\omega)$, which are Fourier transforms of the autocorrelation functions:

$$J(\omega) = \int_{-\infty}^{\infty} G(\tau)e^{i\omega\tau}\,d\tau \tag{63}$$

where ω is the angular frequency. It is clear from equations (56) and (57) that the required functions are just the autocorrelation functions of the Wigner rotation matrices which were defined previously in equation (53) in connection with the fluorescence polarization decay. As in the foregoing discussion these are characterized by exponential decays specified by the correlation times given in equation (40). Again the evaluation of the correlation functions requires knowledge of the orientational pseudopotential, $U(\Omega)$, given for example by equation (43) or (44). The effects of the rotational rates on the spectrum are characterized by the correlation times $\tau_{L,K}$ of equation (40). The principal motional parameters are the rotational diffusion coefficients $D_{R\|}$ and $D_{R\perp}$. Corresponding effective rotational correlation times can be defined if desired, given by $\tau_{R\perp} = 1/6D_{R\perp} = \tau_{20}$ and $\tau_{R\|} = 1/6D_{R\|} = 2\tau_{20}\tau_{22}/(3\tau_{20} - \tau_{22})$.

In the case of slow molecular rotation (most frequent in ESR), time-dependent perturbation theory cannot be applied. Line positions are then not independent of the rotation rates and the full spectral line shape must be calculated. This is done by combining the equation of motion for the magnetic spins with the rotational diffusion equation to give the so-called stochastic Liouville equation (see, e.g., Freed, 1976). The equation of motion for the spins can be expressed in terms of the spin density matrix, ρ (see, e.g., Slichter, 1978):

$$\frac{\partial \rho}{\partial t} = -i[\mathcal{H}(t), \rho] \tag{64}$$

where $\mathcal{H}(t)$ is the total time-dependent spin Hamiltonian including the interaction with the time-varying field which induces the magnetic resonance transitions, and the square brackets denote the commutator: $[A, B] = AB - BA$. Combination with the rotational diffusion equation leads to the stochastic Liouville equation in which the time dependence is then included solely in the elements of the spin density matrix (Kubo, 1969):

$$\frac{\partial \rho(\Omega, t)}{\partial t} = -i[\mathcal{H}(\Omega), \rho(\Omega, t)] - \Gamma_\Omega \rho(\Omega, t) \tag{65}$$

where the operator Γ_Ω is defined by the rotational diffusion equation, which is written as $\partial p(\Omega, t)/\partial t = -\Gamma_\Omega p(\Omega, t)$, i.e.,

$$\Gamma_\Omega = \mathbf{L} \cdot \mathbf{D}_R \cdot \mathbf{L} + \mathbf{L} \cdot \mathbf{D}_R \cdot \mathbf{L} U(\Omega)/kT \tag{66}$$

[cf. equation (41)]. A solution is sought for the departures of the density matrix from its equilibrium value, ρ_0, which has a time dependence of the form $e^{i\omega t}$, where ω is the angular frequency of the time-varying magnetic field which induces the transitions. The solution is obtained by expansion in terms of the eigenfunctions of the diffusion operator, $\psi^L_{K,M}(\Omega) = [(2L + 1)/8\pi^2]^{1/2} \mathcal{D}^L_{K,M}(\Omega)$, whose eigenvalues are $\tau^{-1}_{L,K}$ given by equation (40). This leads to a series of coupled linear equations that must be solved numerically. Again the effects of the rotational motion are characterized by the correlation times $\tau_{L,K}$ of equation (40), the values of which are obtained by comparison of the total simulated line shape with the experimental spectra. Details of the calculation are given in Freed (1976).

Similar methods may be used for the analysis of saturation transfer ESR. In this case the modulation of the static magnetic field must also be included in the time-dependent Hamiltonian in equation (65); see, e.g., Dalton (1985).

3.5. Hydrodynamic Theory

The rotational diffusion coefficient for membrane proteins can be calculated using hydrodynamic theory as was done for the translational diffusion coefficient. In this case the two-dimensional nature of the system does not cause problems and the rotational drag arises from the bulk viscosity of the membrane, η_0. If the viscosity of the aqueous phases can be neglected in the model given in Fig. 5, the friction coefficient for unaxial rotation about the cylinder axis is given by (Saffmann, 1976)

$$f_{R\parallel} = 4\pi\eta_0 a^2 h \qquad (67)$$

The rotational diffusion coefficient is then given again simply by the Einstein relation:

$$D_{R\parallel} = kT/f_{R\parallel} \qquad (68)$$

Hughes *et al.* (1981) have extended Saffmann's analysis to the case in which the viscosity of the aqueous phase is nonvanishing. Their asymptotic result for $\varepsilon = (\eta_1' + \eta_2')a/\eta_0 h \gtrsim 2$ is

$$f_{R\parallel} = 4\pi(\eta_1' + \eta_2')a^3[8/3\pi + 1/\varepsilon] \qquad (69)$$

The analysis for general ε can be found in the same reference. This formulation is of particular interest since it can be applied to proteins that penetrate only partially into the membrane, and thus experience a much larger effective value of η_2'.

A quantity of considerable interest is the ratio of the translational to the rotational diffusion coefficient for a particular protein–membrane system. For the Saffmann (1976) model this is given from equations (22) and (67) by

$$D_T/D_{R\parallel} = [\ln(\eta_0 h/\eta'a) - \gamma]a^2 \qquad (70)$$

If required, the membrane viscosity, η_0, in the logarithmic term can be expressed in terms of the rotational diffusion coefficient, $D_{R\parallel}$. In this way Peters and Cherry (1982) obtained an effective diameter for the bacteriorhodopsin molecule of 43 ± 5 Å from measurement of the translational and rotational diffusion coefficients in dimyristoyl phosphatidylcholine bilayers in the fluid phase. For comparison the maximum cross-sectional diameter of bacteriorhodopsin obtained from structural measurements is 35 Å (Henderson and Unwin, 1975). The effective membrane viscosity of 30°C, obtained from the translational diffusion coefficient at high lipid/protein ratio was $\eta_0 = 1.1$ poise, whereas that deduced from the rotational diffusion coefficient was 3.7 ± 1.3 poise (Cherry and Godfrey, 1981).

ACKNOWLEDGMENT. I would like to thank Dr. Martin D. King for his comments on the manuscript.

REFERENCES

Axelrod, D., Koppel, D. E., Schlessinger, J., Elson, E., and Webb, W. W., 1976, *Biophys. J.* **16:**1055-1069.

Carslaw, H. S., and Jaeger, J. C., 1959, *Conduction of Heat in Solids*, 2nd Ed., p. 334, Oxford University Press, Oxford.

Cherry, R. J., and Godfrey, R. E., 1981, *Biophys. J.* **36:**257-276.

Clegg, R. M. and Vaz, W. L. C., 1985, in: *Progress in Protein-Lipid Interactions* (A. Watts and J. J. H. H. M. De Pont, eds.), Vol. 1, pp. 173-229, Elsevier, Amsterdam.

Cohen, M. H., and Turnbull, D., 1959, *J. Chem. Phys.* **31:**1164-1169.

Dalton, L. R., 1985, *EPR and Advanced EPR Studies of Biological Systems*, CRC Press, Boca Raton.

Devaux, P. F., Scandella, C. J., and McConnell, H. M., 1978, *J. Magn. Reson.* **9:**474-485.

Ehrenberg, M., and Rigler, R., 1972, *Chem. Phys. Lett.* **14:**539-544.

Favro, L. D., 1965, in: *Fluctuation Phenomena in Solids* (R. E. Burgess, ed.), pp. 79-101, Academic Press, New York.

Freed, J. H., 1976, in: *Spin-Labelling, Theory and Applications* (L. J. Berliner, ed.), Vol. I, pp. 53-132, Academic Press, New York.

Galla, H.-J., Hartmann, W., Theilen, U., and Sackmann, E., 1979, *J. Membrane Biol.* **48:**215-236.

Henderson, R., and Unwin, P. N. T., 1975, *Nature* **257:**28-32.

Hughes, B. D., Pailthorpe, B. A., and White, L. R., 1981, *J. Fluid Mech.* **110:**349-372.

Kinosita, K., Kawato, S., and Ikegami, A., 1977, *Biophys. J.* **20:**289-305.

Kubo, R., 1969, in: *Stochastic Processes in Chemical Physics, Advances in Chemical Physics* (K. E. Shuler, ed.), pp. 101-127, Wiley, New York.

Lipari, G., and Szabo, A., 1980, *Biophys. J.* **30:**489-506.

Macedo, P. B., and Litovitz, T. A., 1965, *J. Chem. Phys.* **42:**245-256.

Maier, W., and Saupe, A., 1959, *Z. Naturforsch. A* **14:**882-889.

Marsh, D., 1985, in: *Spectroscopy and the Dynamics of Molecular Biological Systems* (P. M. Bayley and R. E. Dale, eds.), pp. 209-238, Academic Press, London.

van der Meer, W., Pottel, H., Herreman, W., Ameloot, M., Hendrickx, H., and Schröder, H., 1984, *Biophys. J.* **46:**515-523.

Peters, R., and Cherry, R. J., 1982, *Proc. Natl. Acad. Sci. U.S.A.* **79:**4317-4321.

Razi Naqvi, K., 1974, *Chem. Phys. Lett.* **28:**280-284.

Saffmann, P. G., 1976, *J. Fluid Mech.* **73:**593-602.

Seelig, J., 1977, *Q. Rev. Biophys.* **10:**353-418.

Seelig, J., 1978, *Biochim. Biophys. Acta* **515:**105-140.

Slichter, C. P., 1978, *Principles of Magnetic Resonance*, 2nd Ed., Springer-Verlag, Berlin.

Träuble, H., and Sackmann, E., 1972, *J. Am. Chem. Soc.* **94:**4499-4510.

Vaz, W. L. C., Clegg, R. M., and Hallmann, D., 1985, *Biochemistry* **24:**781-786.

Wiegel, F. W., 1980, *Lecture Notes in Physics* (J. Ehlers *et al.*, eds.), Vol. 121, Springer-Verlag, Berlin.

Zannoni, C., 1981, *Mol. Phys.* **42:**1303-1320.

Chapter 5

The Acetylcholine Receptor and its Membrane Environment

F. J. Barrantes

1. INTRODUCTION

The nicotinic acetylcholine receptor (AChR) is by far the best characterized neurotransmitter receptor protein. Various disciplines have contributed
to our current understanding of this macromolecule: Early biochemical
studies were catalyzed by the availability of tissue sources rich in AChR
and appropriate ligands like the α-toxins; more recently the application of
immunochemistry, cDNA recombinant techniques, rapid kinetic methods
and patch-clamp electrophysiological techniques have produced spectacular
advances in the field. The AChR is a glycophosphoprotein of about a quarter
of a million molecular weight composed of two quasi-identical subunits
(α) and three additional chains (β, γ, δ) in a mole ratio of 2:1:1:1
(Reynolds and Karlin, 1978; Raftery et al., 1980; Lindstrøm et al., 1982).
Apparent molecular weights determined by SDS gel electrophoresis are
39,000, 48,000, 58,000, and 64,000, respectively, whereas the *exact* molecular
weights predicted from the nucleotide sequences of the cDNAs coding for
each subunit are 50,166 (α), 53,681 (β), 56,279 (γ), and 57,565 (δ) (Numa
et al., 1983). The α subunits carry the recognition site for agonist and
competitive antagonists. All subunits are glyco-polypeptides and traverse
the membrane at least once.

F. J. BARRANTES • Instituto de Investigaciones Bioquímicas (INIBIBB), Universidad
Nacional del Sur, C.C. 857, 8000 Bahía Blanca, Argentina.

In this chapter I shall review some structural and dynamic properties of the AChR in the membrane environment. Special emphasis will be given to the properties of the lipid bilayer in which the AChR is inserted. The student is referred to more comprehensive reviews (Conti-Tronconi and Raftery, 1982; Changeux et al., 1983; Barrantes, 1983; Popot and Changeux, 1984) for several aspects of AChR structure and function not covered in the present work.

1.1. Structure of the AChR Macromolecule

The detergent-solubilized AChR monomer is an asymmetric cylindrical-shaped body, with a Stokes radius of 70 nm, a radius of gyration of about 4.6 nm, a sedimentation coefficient of 9S, and molecular weight ranging between 290,000 and 303,000 (see references in Barrantes, 1983). The molecular weights more recently determined by radiation inactivation methods (Lo et al., 1982) and Doppler effect laser light scattering (Doster et al., 1980) are close to the 270,000 figure calculated from the primary structure data obtained by cDNA recombinant techniques (see above). In agreement with its subunit stoichiometry (Lindstrøm et al., 1982), the AChR monomer appears to lack symmetry when observed in the *native* membrane (Kistler et al., 1982; Zingsheim et al., 1982a,b). Application of low-dose ($<10e^-/\text{Å}^2$) scanning transmission electron microscopy (STEM) and image averaging techniques specially developed to analyze noncrystalline specimens enabled us to produce the first low-resolution maps of the average AChR particle in the native receptor-rich membranes at a resolution of 1.5–2.0 nm (Zingsheim et al., 1980). The AChR monomer appears in the end-on views as ring-shaped, rosettelike particles 8–9 nm in diameter with a stain-filled central pit. Three predominant nonsymmetrical regions could be identified around the central pit of the AChR in such end-on views.

In the axis normal to the membrane surface, the AChR protein is also unevenly distributed with respect to the bilayer lipid. The mass of the AChR molecule extends 5.5–7.0 nm towards the extracellular space and 1.5–4.0 nm into the cytoplasmic compartment (Klymkowsky and Stroud, 1979; Zingsheim et al., 1982a; Brisson and Unwin, 1985). Upon removal of nonreceptor proteins from the cytoplasmic face of the membrane, a different profile of the AChR particle becomes visible (Barrantes, 1982b). This is presumably the cytoplasmic-facing end of the molecule. Thus, the AChR monomer displays asymmetry along axes both parallel and perpendicular to the membrane plane.

More recently, Brisson and Unwin (1985) have been able to obtain three-dimensional maps of the AChR by Fourier synthesis of images from "annealed" tubular structures derived from AChR membranes. These crystals were of sufficiently good order to perform reconstructions (1.8–2.0 nm resolution) in three dimensions of frozen, unstained, and stained specimens. The AChR appears in these reconstructions as a body with a calculated volume of 350,000 $\mathring{A}^3$, of which only a minor portion is embedded in the membrane. The portion of the AChR protruding about 7 nm into the synaptic cleft displays a fivefold axis of symmetry in a plane perpendicular to the membrane; no symmetric features could be observed in the cytoplasmic portion of the protein, and symmetry could be ascertained in only part of the membrane-spanning domain of the AChR. Given the relative contributions of these three portions, the overall two-dimensional projection of the molecule on the plane of the membrane displays approximate pentagonal symmetry.

Symmetric bodies normally pack in two or three dimensions into regular arrays. The AChR exists in the native membrane as closely but not maximally packed particles (Barrantes *et al.*, 1980; Zingsheim *et al.*, 1980). Ordered arrays (Brisson, 1980; Klymkowsky and Stroud, 1979) occur exceptionally in the native membranes (another reflection of the lack of symmetry of the individual molecule), and the type of symmetry displayed in such ordered lattices is variable (see review in Barrantes, 1979). The usual case is that of disordered particles, although rows of particles have been observed by rapid freezing (Heuser and Salpeter, 1979) and negative contrast techniques (Barrantes, 1982b). Using a combination of (1) depletion of nonreceptor proteins by alkaline treatment and (2) a mild physical shearing of the resulting fragile membranes, oligomeric forms of the AChR could be observed in the membrane by negative contrast electron microscopy (Barrantes, 1982b). The 9S AChR monomer predominant in reduced membranes, the 13S species in normal or alkylated membranes, and higher oligomeric forms, both in normal or oxidized membranes, could be correlated with the AChR species revealed by gradient centrifugation and with the presence or absence of some nonreceptor proteins (Barrantes, 1982a,b).

1.2. Subunit Localization in the AChR Macromolecule

Lactoperoxidase iodination (St. John *et al.*, 1982), graded proteolysis in sealed, intact vesicles (Strader and Raftery, 1980), chemical analysis of AChR chains, and cross-linking with neurotoxins (see references in Conti-Tronconi and Raftery, 1982) indicate exposure of all subunits to the extracellular millieu. Secondly, antisubunit antibodies (see Lindstrøm *et al.*, 1983) also make clear the exposure of all subunits to the intracellular space (Froehner, 1981). Thirdly, the intramembranous portion of AChR subunits can be reached from the bilayer phase with lipophilic reagents like pyrene sulphonyl-azide (Sator *et al.*, 1979) or iodo-naphtyl-azide (Tarrab-Hazdai *et al.*, 1980). Hidden antigenic determinants of the AChR can be exposed after depletion of peripheral, nonreceptor proteins (Froehner, 1981). Some of these proteins, in turn, have been located at the cytoplasmic face of the membrane (Barrantes, 1982b; St. John et al., 1982; Nghiem et al., 1983; Sealock et al., 1984). Taking these pieces of evidence together, it becomes apparent that all subunits are exposed to the extracellular and cytoplasmic compartments and traverse the membrane at least once. In fact (see Section 4.1), current models postulate the existence of several transmembrane helices for each chain, bringing the total number of helices within the bilayer to more than 25. The AChR is an *integral, transmembrane glycoprotein.*

Direct structural studies aimed at locating individual subunits have also been conducted. In addition to the 9S AChR monomer, a second 13S form with a Stokes radius of about 8.5 nm and a molecular weight twice that of the monomer is also present in AChR membranes. Interconversion of the two forms occurs via disulfide reduction–oxidation. The comparison of native membranes with those subjected to cleavage of the disulfide bonds by reducing agents has yielded the putative average position of the δ subunits (Zingsheim et al., 1982a) in the native membrane-bound AChR. Comparable results were obtained by measurement of the angles determined by biotinylated toxin-avidin complexes tagging α subunits in AChR trimers (Wise et al., 1981). Similarly, Fairclough et al. (1983) measured the angles subtended by Fab fragments of monoclonal antibodies directed against the α subunits on AChR particles. They concluded that the δ–δ bonded dimers are a mixed population of translationally related particles (R–R) and particles related by a C2 type of symmetry (R–Я). The angular distributions of Fab fragments are broadened by the contribution arising from the flexibility about the δ–δ bond, which was calculated to be ±22°. The reason for the different occurrence of symmetrical dimers related by a twofold axis in native membranes (Zingsheim et al., 1982a) and asymmetrical dimers in lipid-enriched or reconstituted vesicles (Fairclough et al., 1983); Bon et al., 1984) is not known. The nature of the interactions operative in the two types of preparation might bear on the discrepancy.

In spite of the ambiguities mentioned in the preceding section, knowledge of the location of the δ–δ bond linking two monomers in a dimer provides a useful landmark for establishing the relative topography of other AChR subunits in the population of dimers displaying C2 symmetry in the native AChR membranes (Zingsheim et al., 1982a). Since the α subunits carry the recognition site for agonists and antagonists, the latter can be used to locate these chains by structural methods. The use of avidin bound to biotinylated α-toxin enabled Holtzman et al. (1982) to calculate the angles separating the α subunits. The conclusion was reached that the two α chains in a monomer cannot be contiguous. The angle between α subunits (110 ± 30°) was also measured with avidin–biotin in artificially produced dimers cross-linked via the β subunits (Karlin et al., 1983). We have made use of native α-bungarotoxin for the direct location of its recognition sites on the AChR molecule by low-dose electron microscopy and single-particle image-averaging techniques (Zingsheim *et al.*, 1982b). Attachment of a single native toxin to its recognition site on each α chain produces a significant increase in the mass contributing to the average image of the AChR, roughly one-quarter of the mass per α subunit. A difference map between the toxin-tagged and untreated membranes yields two statistically significant peaks of stain-excluding density. The two α chains exhibit contacts with a different set of subunits, i.e., they possess different local environments. This provides a structural framework within which one can rationalize the functional nonequivalence of the recognition sites.

Fairclough *et al.* (1983) have employed immunocytochemistry to measure the angle between α subunits, which was found to be $144 \pm 4°$, i.e., nonadjacent within the monomer. Small-angle X-ray diffraction of toxin-tagged AChR also suggested an apical location of the binding sites on the extracellular portion of the molecule (Fairclough *et al.*, 1983). Bon *et al.* (1984) have also carried out an analysis of membrane-bound AChR tagged with native α-bungarotoxin; their results are in full agreement with ours (Zingsheim *et al.*, 1982b) on the location of α subunits. The nonadjacent distribution of α subunits is corroborated by recent biochemical studies (Hamilton *et al.*, 1985). Lindstrøm *et al.* (1983) quote results indicating that only one of the two α chains is glycosylated, a further manifestation of asymmetry in the AChR, in this case at the level of individual subunits within the receptor monomer. A more recent report on this issue (Conti-Tronconi *et al.*, 1984) is still open to controversy.

2. AChR PRIMARY STRUCTURE, cDNA RECOMBINANT TECHNIQUES, AND AChR MODELS

2.1. Common Features of AChR Subunits

Our understanding of AChR structure advanced considerably upon application of modern molecular biology methods such as cDNA recombinant techniques. This advance was based on the finding, obtained with conventional chemical sequencing methods, of the sequence of the first 54 amino acid residues from the NH-terminus of the four subunits (Raftery *et al.*, 1980). The sequence homologies found between the aminotermini of the subunits suggested the evolutionary origin of all chains from a common ancestral gene, followed by divergence through gene duplication (Raftery *et al.*, 1980). Calculations of the secondary structure of the AChR subunits emerging from the above studies have also been attempted (Guy, 1981). The information obtained from such limited sequence data indicates considerable α-helical and β-pleated structure in all subunits.

Using the known limited sequences as a starter, cDNA recombinant techniques could be applied to determine the partial (Ballivet *et al.*, 1982; Sumikawa *et al.*, 1982) and subsequently the full nucleotide sequences of all *Torpedinidae* AChR subunits (Noda *et al.*, 1982, 1983a,b,c; Claudio *et al.*, 1983; Devillers-Thiery *et al.*, 1983) and hence the complete primary structure of the AChR chains was rapidly elucidated. This knowledge accelerated the study of the corresponding subunits of calf AChR (Noda *et al.*, 1983a; Takai *et al.*, 1984; Tanabe *et al.*, 1984; Kubo *et al.*, 1985) and the α and γ subunits of human muscle AChR (Noda *et al.*, 1983a; Shibahara *et al.*, 1985). Comparison of the amino acid sequences (Noda *et al.*, 1983c; Numa *et al.*, 1983) shows that the homology between chains extends through most of the primary structure of the subunits with the exception of small stretches, particularly between residues 342 and 425. This is indicative of the existence of similar structural features in each subunit and hence of similar contributions of each subunit to the overall AChR structure. It also reinforces the view (Raftery *et al.*, 1980) that the subunits could have evolved from a common ancestral gene. Another interesting offspring of comparative studies is the unexpected finding of a fifth subunit in calf AChR (Takai *et al.*, 1985) which shows higher homology with the γ chain than with the others. This subunit, termed ε, is only present in the adult muscle, a fact that may help explain various biologically relevant differences between adult and less than fully developed AChR.

The distribution of amino acid residues along subunit polypeptide chains can be compared and also analyzed in terms of the occurrence of hydrophobic and hydrophilic stretches. This is most conveniently done by application of algorithms such as those developed by Kyte and Doolittle (1982) or Hopp and Woods (1981). This type of analysis makes apparent the occurrence of four segments of hydrophobic amino acid residues common to all receptor subunits (Noda *et al.*, 1983a,b; Fairclough *et al.*, 1983; Finer-Moore and Stroud, 1984).

2.2. AChR Subunit Transmembrane Portions: Odd versus Even Number of Crossings

The remarkable alignment of similar stretches along the chains, which brings into register the hydrophobic regions of the different subunits, led Noda *et al.* (1982), Claudio *et al.* (1983), and Devillers-Thiery *et al.* (1983) to postulate models of the subunit arrangement with respect to the lipid bilayer in which each of the four hydrophobic domains traverses the membrane at least once and displays α-helical configuration. These postulations were based on (1) the observed length of the hydrophobic stretches; (2) the analogy with other transmembrane proteins like bacteriorhodopsin and glycophorin A, and (3) favorable length of the observed stretches in α helix as opposed to other polypeptide conformations (e.g., β sheets) to cover the bilayer thickness. These models predict that the amino acid and carboxyl termini of the subunits face the extracellular millieu. This is in contrast with the data obtained on other integral membrane proteins studied to date, in which the N-terminus is expected to remain on the extracellular space after cleavage of the leader sequence, and the C-terminus is exposed to the intracellular face of the membrane (Sabatini *et al.*, 1982). In fact, available experimental evidence points to the location of the N-terminus of the δ subunit on the extracellular face of the membrane (Anderson *et al.*, 1983). This type of model—the so-called four-helix model—implies the presence of about 30% of the AChR protein on the cytoplasmic face of the membrane.

An alternative type of model (the five-helix model) places a much smaller (ca. 20%) portion of the protein on the intracellular face of the membrane and predicts that in addition to the four transmembrane helices per chain there is a fifth, amphipathic transmembrane helix (Fairclough *et al.*, 1983; Guy, 1983). The basic discrepancy between the two types of model is that the five-helix model places the C-terminus on the cytoplasmic face of the membrane. The aforementioned authors also produce evidence suggesting that the amphipathic helix possesses a continuous hydrophobic face on one side and a hydrophilic face on the other, thus providing a structural substratum for the lining of the internal walls of the ionic channel by the latter. The continuous hydrophilic face has the appropriate length to span the lipid bilayer (about 3.8 nm) and a distribution of charges (21 positive, 19 negative, 10 uncharged) in its residues (Fairclough *et al.*, 1983) that makes it ideally suited for intersubunit ion pairing. Hypotheses on the amino acid arrangement of the channel-lining portion of the subunits have appeared (Kosower, 1983, 1984).

2.3. The Synaptic Cleft-Facing Portion of the AChR

The regions between residues 70 and 160 in the four subunits are highly homologous (27 residues are totally conserved). Both the four- and five-helix models postulate that this region is on the synaptic cleft. According to Numa *et al.* (1983) this region would carry the important features: a site for N-glycosylation and the disulfide bond, which could possibly react after reduction with affinity reagents like 4-(*N*-maleimido) benzyltrimethyl-ammonium (MBTA) on the α subunit. More recently Kao *et al.* (1984) have established that Cys 192 and possibly Cys 193 are responsible for MBTA binding (see below). This stretch is also postulated to be rich in amphipathic β sheets common to all subunits in the form of a six- or eight-strand β barrel domain (Finer-Moore and Stroud, 1984). The "solvent" side of the β barrel could line the central well of the receptor. In the α subunit a β bend in this barrel near two putative disulfides established between Cys 128 and 142 could form part of the ACh-binding site as suggested by Numa *et al.* (1983). According to Fairclough *et al.* (1983) the predominant negative charges in the "vestibule" of the receptor might contribute to the cation specificity of the receptor.

2.4. Testing of Models: The Immunochemical Approach

Poly- and monoclonal antibodies of different degrees of specificity (antitotal AChR, antisubunit, etc.) have been successfully used to map regions of the receptor molecule inaccessible to other techniques such as affinity labeling or chemical modification. A main immunogenic region ("MIR," Lindstrøm *et al.*, 1982), consisting of at least two adjacent epitopes close to the ACh recognition site on the α subunit, appears to be a phylogenetically preserved immunological feature of the AChR. Antibodies against the MIR cross-react between species and passively transfer the acute form of EAMG (experimental autoimmune myasthenia gravis). Within the extensive antibody library available to date, some types noncompetitively inhibit the $^{22}Na^+$ flux, and others are able to distinguish between the AChR monomer and dimer or cross-link two δ subunits (Lindstrøm *et al.*, 1982). By using monoclonal antibodies against synthetic peptides prepared on the basis of the known sequence data, Lindstrøm *et al.* (1984) have demonstrated that the C-termini of the AChR chains are exposed to the cytoplasmic face of the membrane, thus favoring the five-helix type of model.

Recent studies of Juillerat *et al.* (1984) and Lindstrøm *et al.* (1984) also served to test other predictions about the putative location of the MIR (Noda *et al.*, 1982) based on the sequence data obtained from the cDNA nucleotide structure. The latter authors had suggested that the most hydrophilic domain (residues 161–166) of the α subunits formed the main immunogenic region. These predictions were based on the knowledge that the most hydrophilic regions of the amino acid sequence in a protein are prominent antigenic determinants (Hopp and Woods, 1981). Lindstrøm *et al.* (1984) subsequently demonstrated that the 161–166 stretch of the α chains is not the MIR and is not even contiguous to this region. Numa *et al.* (1983), however, have alternatively located the MIR in the α 197–207 stretch.

More recently, antibodies raised against synthetic peptides corresponding to defined sequences of the AChR chains have been used to test available models of AChR orientation in the membrane. Thus Criado *et al.* (1985) found that residues 152–159 in the α-chain C-terminus are facing the cytoplasmic side of the membrane in order to account for their findings, a new model with seven transmembrane domains (two additional amphipathic segments) was proposed. Similarly, Young *et al.* (1985) developed polyclonal antibodies against a synthetic hexadecapeptide corresponding to residues 501–576 at the COOH-terminus of the δ subunit and against the 350–358 stretch of the β subunit. They could demonstrate that the C-termini of these two subunits were also located at the cytoplasmic side of the membrane. Finally, LaRochelle *et al.* (1985) have shown the cytoplasmic exposure of the γ 360–370 stretch. In summary, the COOH-terminus of all subunits appears to be located on the cytoplasmic compartment thus favoring the odd-crossing type of model.

Figure 1 summarizes the current views on the organization of the AChR in the postsynaptic membrane. The models are constrained by several factors related to the biochemistry and immunochemistry of the AChR protein. Two of these concern the location of the carbohydrate moiety of the glycoprotein and of the disulfide bonds. It has been proposed that the *N*-glycosidic linkage occurs at Asn-141 (Noda *et al.*, 1982), and hence this residue should be exposed at the extracellular space. This appears to be the case, as recent site-directed mutagenesis of α-chain DNA with replacement of Asn-141 by aspartic acid seems to indicate (Mishina *et al.*, 1985). As can be observed in Fig. 1, all types of model locate the site of *N*-glycosylation on the extracellular space. Continuing from this point, twelve amino acid residues separate Asn-141 from residue 152, which is postulated to be exposed (α 152–159) at the cytoplasmic side (Criado *et al.*, 1985). The 12-amino acid stretch is not sufficiently long to traverse the membrane in an α-helical configuration, for which reason a β conformation was proposed for the M6 segment (Fig. 1). Other similarities and discrepancies between the two types of model are discussed in Fig. 1.

As far as the acetylcholine binding site is concerned, we currently have some indications on which segment of the α subunit may be involved. Numa's group (Noda *et al.*, 1982; Numa *et al.*, 1983) initially postulated that α Cys 128 and Cys 142 participated in the acetylcholine binding site, and the hypothesis was recently tested by point mutation of the residues in the α chain (Mishina *et al.*, 1985). When Cys 128 and Cys 142 are replaced by serine residues (hence precluding disulfide bond formation) antagonist binding and electrophysiological responses are hindered (Mishina *et al.*, 1985; see below). The identification of Cys 192 and possibly Cys 193 as the sites specifically labeled by the affinity reagent MBTA (Kao *et al.*, 1984) also place these two residues on the extracellular side of the membrane and suggest their involvement in the acetylcholine binding site. Furthermore, since immediately adjacent cysteine residues (Cys 192 and 193 in the α chain) are unlikely to form disulfide bonds (for steric reasons), other matching partners must be sought for the bonding. Cys 128 and Cys 142 appear to be the most likely candidates, though no labeling with MBTA could be observed in these two residues (Kao *et al.*, 1984). McCormick and Atassi (1984) could test the ability of α 125–147 to retain cholinergic ligand binding, and more recently Lennon *et al.* (1985) have provided

4 - transmembrane domains

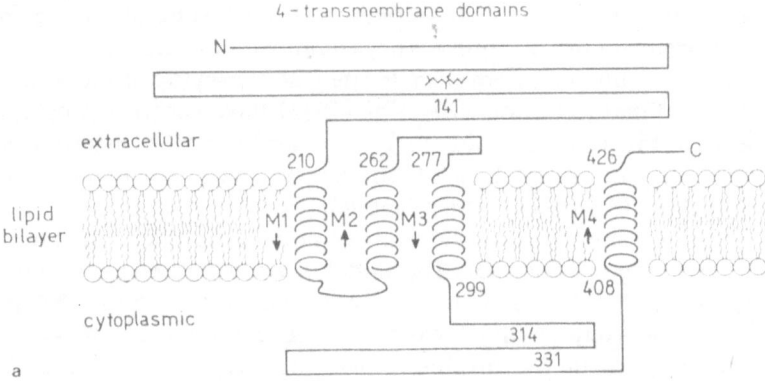

a

5 - transmembrane domains

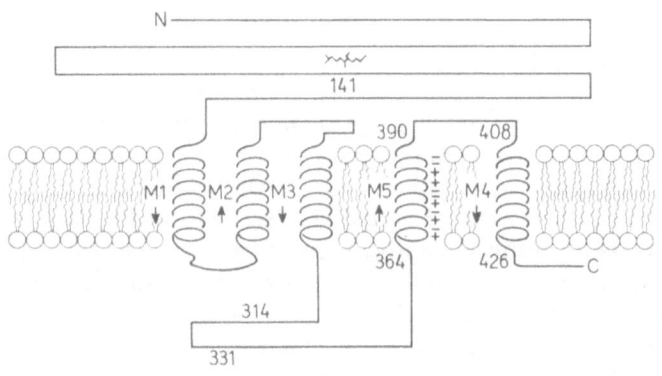

b

7 - transmembrane domains

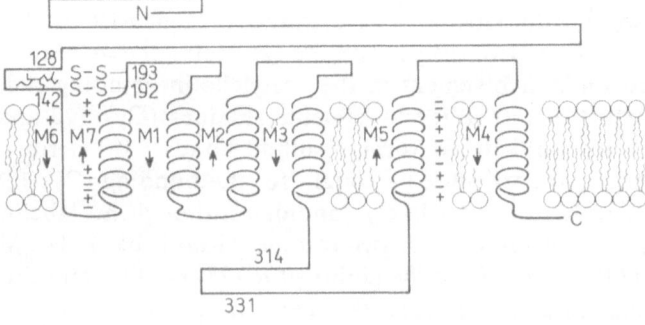

c

immunochemical evidence on the extracellular location of this segment. They also showed that the 125-147 stretch is capable of inducing EAMG, T-cell immunity, and modulating autoantibodies. Wilson *et al.* (1985) have identified the α-toxin binding site within the stretch 153-241 of the α subunit, and the lack of participation of the α 1-140 segment was also reported. More precisely, the segment α 173-204 was proposed as a "major determinant" of the bungarotoxin recognition site (Wilson *et al.*, 1985). The vicinity of this site and the disulfide bond on the α subunit has been reported (Momoi and Lennon, 1986). In summary, a combination of data obtained from cDNA recombinant and sequencing techniques, immunochemistry, immunocytochemistry, and affinity labeling experiments together with theoretical analysis has been used to propose and test models of AChR structure and organization in the membrane. An odd number of subunits traversing the membrane appears to date as the most plausible, albeit not yet proven, model.

3. AChR-MEDIATED CHANNEL GATING: A SINGLE MOLECULE AT WORK

3.1. The 9S Monomer as Minimal Binding/Gating Unit

A most challenging problem, perhaps the ultimate aim in the study of any biomolecule, is to attempt correlation of its structure and function at a sufficiently detailed level. Which is the minimal structure capable of generating a permeability response upon binding of acetylcholine? Which subunits of the AChR and portions thereof participate in the translocation of ions? Which molecular arrangement is necessary for optimal conductance through the channel? This is the type of question that must be asked in addressing the above problem, and only the first is answered to the level of resolution we can afford today. Even the description of the molecular mechanisms leading from agonist binding to gating of the AChR-controlled channel is still open. We are, however, in a favorable position to tackle some of these queries at the level of individual AChR molecules. The success

---

Figure 1. Models of AChR structure. Common to all models is the long hydrophilic extracellular domain at the N-terminus, three closely spaced α-helical transmembrane portions (M_1-M_3), a hydrophylic cytoplasmic domain, and a fourth transmembrane helix (M_4). Dissimilarities comprise the postulation of a fifth amphipatic helic $[M_5$ in (b)], two additional transmembrane domains $(M_6$ and M_7 in (c)], and the extracellular (a) versus cytoplasmic (b, c) location of the C-terminus. Minor differences are the fully exposed (a, b- or hidden (c) location of the N-terminus. The 4-helical model (a) was proposed by Noda *et al.* (1982), Claudio *et al.* (1983), and Devillers-Thiery *et al.* (1983). The 5-helical model was conceived by Fairclough *et al.* (1983), Guy (1983), and Young *et al.* (1985). The 7-transmembrane domains were proposed by Criado *et al.* (1985).

of cDNA recombinant and sequencing techniques in deriving the primary structure of the AChR protein has already been discussed; here I shall comment on a technique that enables the direct observation of the elementary currents associated with ion permeation through a single AChR-controlled channel. This can be accomplished both in the living cell and in artificial, reconstituted membranes by the application of an electrophysiological recording technique, the patch-clamp method (Neher and Sakmann, 1976), and its high (gigaohm) resistance improvements (Hamill *et al.*, 1981). In reconstituted systems (1) the chemical composition of lipid and protein, (2) the thermodynamic state of the host membranes, (3) the degree of purity of the AChR, (4) its oligomeric state, and (5) the sidedness of the membrane can be controlled more easily than in the native membrane (see Latorre *et al.*, 1985 and references therein). One observation that stems from *in vitro* studies is that the M_r 250,000 monomer, with its pentameric structure, is the minimal functional unit possessing ligand recognition ability and controlling the ionic permeability (Boheim *et al.*, 1981). It is also possible to tentatively relate the AChR channel states to corresponding states of ligation of the AChR recognition site (see review in Barrantes, 1986).

One of the questions posed above, namely, which subunits are involved in the permeability response of the AChR to agonist binding, has already begun to be tackled experimentally. Mishina *et al.* (1985) and White *et al.* (1985) have recently reported on site-directed mutagenesis experiments that test the ability of different combinations of AChR subunits to produce functional permeability responses. Deletions and replacements of critical amino acids were performed by Mishina *et al.* (1985) and tested electrophysiologically after injecting the mRNAs in *Xenopus* oocytes. Altered α subunits and normal β, γ, and δ chains abolished ACh-induced depolarizations, indicating the importance of the α chains in the physiological response. These authors were also able to make more precise statements on the location and function of cysteine residues (see above). White *et al.* (1985) used a different approach. They synthesized mRNAs coding for *Torpedo* α, β, and γ subunits and combined these with mouse δ subunit, injecting them into *Xenopus* oocytes. The hybrid was even more effective than the normal message—i.e., augmented ACh binding, larger single-channel conductance, smaller mean channel closed times, and slower rates of desensitization were observed with the hybrid than with the native species.

Sakmann *et al.* (1985) have more recently combined the power of single-channel recording and cDNA recombinant techniques to further define the relationship between AChR subunits and gating behavior of the receptor channel. Hybrids of calf and *Torpedo* AChR subunits were injected in *Xenopus* oocytes and compared with the corresponding native receptors, which exhibited average currents of about 1600 nA and 23 nA, respectively. The difference in macroscopic currents elicited by each type of AChR might be due to the ability of calf AChR to let more ions pass through its channel, or to a higher probability of maintaining the channel open at a given agonist concentration. The possibility that more channels were synthesized by the oocyte with calf message was discarded. Single-channel measurements showed the two channels to differ substantially in their gating behavior: the average duration of elementary currents flowing through *Torpedo* receptors was much shorter (~0.6 msec) than that of calf AChR channels (~8 msec). *Torpedo* receptors also appeared not to be voltage dependent, whereas calf AChR did show voltage sensitivity. In spite of these differences, the two native receptors were similar in ion selectivity and transport properties. Substitution of the calf δ subunit in the *Torpedo* AChR altered its gating behavior, making it similar to the calf AChR, whereas substitution of the calf α subunit was less effective in altering the gating of the *Torpedo* channel. Thus, the mean current duration of the α hybrid (α-chain substitution) is between those of the *Torpedo* and calf native forms, while the δ-substituted hybrids are more similar in duration to the calf wild form. The lapse for which the channel remains open is apparently determined by features present in the α and δ subunits. Selective mutagenesis, and single-channel analysis of protein synthesized with point-altered messages are possible paths for the refinement of this type of data in the near future. Experiments of this sort will also enable the testing of models derived from analysis of sequence data and selective chemical modifications of the protein aimed at deciphering the putative regions of the AChR involved in ligand binding and lining of the ionic channel.

4. THE AChR IN THE LIPID BILAYER

4.1. Theoretical Postulates of AChR Location in the Membrane

The AChR models which we have analyzed in the preceding sections specifically address the question of which portions of the receptor are in contact with the membrane lipid bilayer, a subject which has also been tackled experimentally in the past. Thus, photoreactive probes with a partition coefficient for penetrating favorably the hydrocarbon region of the membrane bilayer have been used to demonstrate crosslinking of these reagents with the α chain (Tarrab-Hazdai *et al.*, 1980), the β and γ subunit (Sator *et al.*, 1979) or all four subunits (Middlemas and Raftery, 1983). Photoreactive phosphatidylcholine analogs have been employed more recently, also resulting in the labeling of all subunits (Giraudat *et al.*, 1985). This lends support to a common feature of all theoretical models referred to above, i.e., their proposal of transmembrane portions exposed to the lipid bilayer for all AChR subunits.

Although the interactions between lipids and AChR protein must be extensive and mutual, little is known about their nature and consequences on receptor structure and function. The study of the influence of the lipid environment on AChR properties is being tackled in our laboratory at various levels of complexity and with different aims in mind: (1) lipid composition and metabolism of the electrocyte *in vivo* in order to understand the overall characteristics of the lipid millieu and the time domains of its turnover; (2) lipid metabolism *in vitro* using an electrocyte preparation to gain a closer insight into the kinetics of lipid incorporation in a closed system; (3) lipid composition and protein–lipid interactions in isolated AChR-rich membranes, in order to define the immediate microenvironment of the AChR and the essential lipid requirements for its function; and (4) studies on reconstituted model systems in which the contribution of individual lipid classes on AChR channel gating function can be separately assessed.

4.2. Composition of the *Torpedinidae* Electrocyte and AChR Membranes

A comparison of the lipid composition of three *Torpedinidae* species (*Torpedo marmorata*, *Torpedo californica*, and *Discopyge tschudii*) has been carried out (Rotstein *et al.*, 1987b). Qualitative differences are apparent especially in their fatty acid composition, which varies among fish species. Monoenoic fatty acids, saturates, and 20 and 22 carbon polyenes are the major acyl chains of glycerophospholipids and neutral lipids in the electrocyte. Compositional studies of electrocyte phospholipids reveal heterogeneity of molecular species, which is evidenced also in the separation of classes differing in physical properties. Such differences might also occur in the membrane, giving rise to separated lipid domains having distinct bulk physical properties. As far as fish species is concerned, docosahexaenoate was the major polyene in all cases, docosapentaenoate (*n*-3) was more abundant in *D. tschudii* lipids, and 20-carbon polyenes were more abundant in *T. californica* lipids than in the other two species (Fig. 2). Differences are less apparent when phospholipid classes of AChR membranes are compared (Table I). Taking all phospholipid classes together, about half the fatty acids in AChR membranes (48%–58% in the three *Torpedinidae* species) are long-chain polyunsaturated fatty acids, in agreement with reports on single species (Popot *et al.*, 1978; González-Ros *et al.*, 1982). Docosahexaenoic acid makes up 70% of the latter in *T. marmorata*, 40% in *D. tschudii*, and 55% in *T. californica* (Fig. 2).

In contrast with all other phospholipids, phosphatidylcholine (PC), the major phospholipid class in AChR membranes (Table I), is made up of 72% monoenoic and disaturated fatty acid species and only 27% of polyunsaturated fatty acids (Fig. 2). In phosphatidylethanolamine (PE) and phosphatidylserine (PS) polyenoic species represent 77% and 57%, respec-

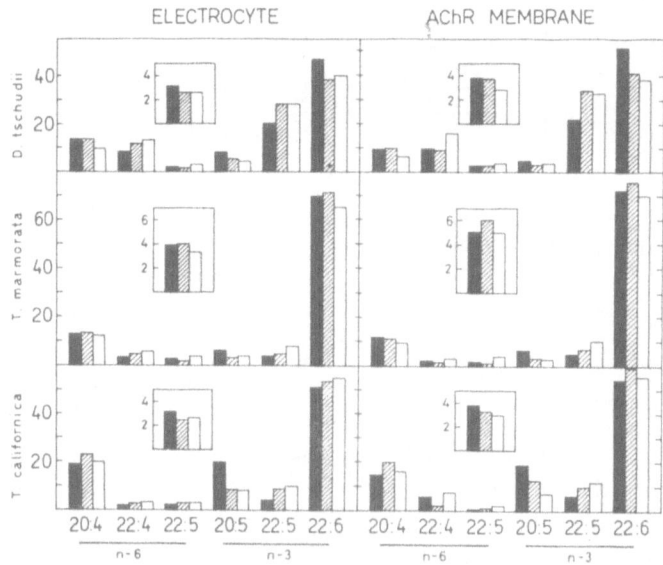

Figure 2. Molar ratios between 20- and 22-carbon *n*-6 and *n*-3 polyunsaturated fatty acids in phosphatidylcholine (■), ethanolamine phosphoglycerides (▨), and phosphatidylserine (□) in the whole electrocyte (left panels) and AChR membranes of three *Torpedinidae* (right panels). Only the depicted fatty acids are considered in the calculations. Data on *T. californica* AChR membranes were calculated from the work of González-Ros *et al.* (1982). *Insets:* Ratio between the sum of *n*-3 and the sum of *n*-6 fatty acids. From Rotstein *et al.* (1987b).

tively. Hexaenoic molecular species constitute the major type of polyenoic fatty acids in PE and PS, a very small proportion being dipolyunsaturated. Saturated species are present in lower proportion than monoenes in PE and PS, in contrast with PC (Fig. 2).

The content of polyenoic fatty acids other than arachidonate in phosphatidylinositol of AChR membranes is higher than usual. Sphingomyelin also shows a peculiar fatty acid composition: it displays a high proportion of short-chain (~14 C) fatty acids, a large content of 22:1, and very little 18:0 in comparison with mammalian nervous tissue sphingomyelins.

In an attempt to better characterize the lipid environment of the AChR protein, serial extractions were carried out on AChR membranes with solvents of increasing polarity (Fig. 3). Hexane, a nondenaturing organic solvent, extracts roughly equal amounts of choline- and ethanolamine-phosphoglycerides, leaving the AChR protein in the membrane residue. Subsequent extractions with chloroform:methanol mixtures of increasing polarity release proportionally lower amounts of PE and concomitantly change the phosphatidylcholine:phosphatidylethanolamine ratio in favor of the former. Thus, as judged by this criterion the phospholipid most

Table I. Comparative Phospholipid Composition of Acetylcholine Receptor Membranes of *Torpedinidae* Electric Organs[a,b]

	Torpedo marmorata				Discopyge tschudii[b]	Torpedo californica		Torpedo oscellata[h]
	b	c	d	e		f	g	
PC	41.2 ± 1.51	38.8 ± 0.3	46	40.7	44.2	40.6	38.4	27.9–42.3
PE	33.2 ± 1.51	38.6 ± 2.1	31	34.8	32.1	29.1	42.2	33.1–44.0
PS	15.0 ± 0.14	16.9 ± 1.1	14	6.9	12.5	10.7	10.9	10.1–15.1
PI	3.6 ± 0.08		2 ± 1		1.9	1.7	N.F.	1.21–1.88
PIP	0.13	—	—	—	0.14	—	N.F.	—
PIP$_2$	0.14	—	—	—	0.04	—	N.F.	—
SPH	5.1 ± 0.48	1.1	7	8.0	8.0	9.3	2.3	1.41–1.93
PA	0.2	—	—	—	0.4	0.9	2.9	—
CL	1.5 ± 0.05	—	—	9.5	0.7	0.9	3.3	1.03–1.93
Phospholipid/cholesterol ratio (i)	1.73 ± 0.15	1.1–1.18	3.5	0.91 ± 0.2	1.30	—	1.1 ± 0.27	—
Lipid/protein ratio	0.69 ± 0.08	0.45–0.67	0.43	0.48	0.62	—	0.43 ± 0.12	—

[a] Expressed as percentage of total lipid phosphorus. N.F., expressly stated as not being found.
[b] This work. The specific activity of the membrane fractions, reflecting their enrichment in AChR protein was 1200 pmole α-bungarotoxin per milligram of membrane protein and 1.000 pmole/mg protein for *T. marmorata* and *D. tschudii*, respectively.
[c] Data from Popot et al. (1978). PS and PI are summed up. Specific activity of the membranes ranged between 3.000 and 4.500 pmole/mg protein.
[d] Data from Schiebler and Hucho (1978). PS and PI are summed up. The same applies to PA + CL. Specific activities: 2.000–4.000 pmole/mg.
[e] Data from Lantz et al. (1985).
[f] Data from Michelson and Raftery (1974).
[g] Data from González-Ros et al. (1982). Specific activity was 1.500 pmole/mg protein.
[h] Data from Kostic et al. (1972).
[i] The phospholipid/cholesterol ratio is expressed in moles per mole.

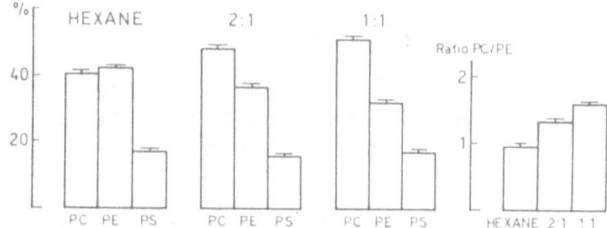

Figure 3. Percentage distribution of the major phospholipids sequentially extracted from AChR membranes with solvents of increasing polarity: hexane and chloroform : methanol (2 : 1 and 1 : 1 v/v) (N. Rotstein and F. J. Barrantes, unpublished).

tenaciously associated with the AChR protein appears to be phosphatidylcholine (Fig. 3). This is noteworthy, given the unique characteristics of this phospholipid in terms of fatty acid content (~50% saturates) in comparison with all other phospholipids in the membrane.

4.3. Lipid Metabolism in Whole Electric Organ and Living Electrocytes

Studies on lipid metabolism in electric tissue *in vivo* are rather scarce. The only previous reports on the subject have dealt with the metabolic fate of intraperitoneally injected ^{32}P-phosphate (Trams and Brown, 1972) and ^{32}P-glucose-6-phosphate after direct injection into the electric organ (Bleasdale *et al.*, 1976). *In vitro*, the incorporation of ^{32}P-phosphate and U-^{14}C-glucose has been measured in slices of *Electrophorus electricus* Sach's organ (Rosenberg, 1973). The incorporation and distribution of [^{3}H]- and [^{14}C]-labeled glycerol and two fatty acids, [^{3}H]oleate (18:1) and [^{3}H]arachidonate (20:4), into lipids of *Discopyge tschudii* electrocytes was followed after direct injection into the electric organ and after incubation of electrocyte stacks with the precursors (Rotstein *et al.*, 1987a). The incorporation of both fatty acids in electrocyte lipids *in vitro* was highly efficient and sustained, attaining similar distributions among phospholipids as *in vivo*. The results indicate that the electrocyte actively synthesizes its own lipids *de novo* and that deacylation–reacylation processes are operative in this cell, unsaturated acyl moieties undergoing a fast turnover. Enzymes catalyzing these reactions may play important roles in the control of the physical properties of electrocyte membranes that depend on lipid class proportions and on the length and unsaturation of lipid acyl moieties. In particular, the fatty acid experiments indicate the presence of acylCoA-lysophospholipid acyl transferases in the electrocyte. These enzymes carry out replacement, exchange, and rearrangement of membrane–lipid acyl chains, processes that have been proposed to be involved in the control of membrane physical properties (such as "viscosity") that depend on the quality of lipid acyl groups (Bell and Coleman, 1980). The *in situ* "retailoring" of lipid molecular species mediated by these processes allows for rapid adjustment of membranes to abrupt changes in environmental conditions, whereas long-term adaptation is controlled by modifications of the *de novo* lipid synthetic processes (Lynch and Thompson, 1984).

4.4. AChR Lipid Interactions: Long-Chain Phospholipids and Cholesterol

Many of the initial studies on AChR reconstituted into artificial lipid systems have employed asolectin, a very complex mixture that includes most common phospholipids as well as 5%–20% lysophospholipids and steroids (Verstraete and Vandenbroucke, 1956; Letters, 1964). One of the first attempts to establish the influence of lipids on AChR properties made use of simpler lipid mixtures (Kilian *et al.*, 1980). The extent of agonist-stimulated sodium flux was the parameter measured in this case. The authors observed that vesicles reconstituted with asolectin that was depleted of neutral lipids, or with mixtures of pure phospholipids, exhibited smaller $^{22}Na^+$ influx than those obtained with complete asolectin. Inclusion of several neutral lipids (α-tocopherol and some quinones) had the reverse effect, provided phosphatidylserine was present. Dalziel *et al.* (1980) and McNamee *et al.* (1982) subsequently reported an increased receptor-mediated ion influx into phosphatidylethanolamine/phosphatidylserine vesicles when their cholesterol content was raised. These results can be correlated with the preferential affinity for cholesterol exhibited by the purified AChR in monolayer systems (Popot *et al.*, 1978) and the importance of this sterol in maintaining the appropriate cohesive pressure for the AChR to function in a bilayer (Schindler, 1982). We have found that not only is cholesterol necessary for the preservation of the ion flux after reconstitution, but it is also essential for the preservation of the agonist-induced affinity transitions (Criado *et al.*, 1982a). Furthermore, we found different cholesterol requirements for the detergent solubilization and subsequent reinsertion of the AChR in a bilayer. Omission of these considerations can give rise to reconstituted systems with irreversibly altered (>95%) receptor (e.g., Lüdi *et al.*, 1983). We have subsequently reported optimal concentrations of cholesteryl esters and synthetic phospholipids for preservation of *both* ligand binding and permeability responses after reconstitution of AChR into liposomes (Criado *et al.*, 1984).

In spite of the biochemical evidence accumulated over the past years on the importance of steroids in AChR function, there are still discrepancies on the topographical location of these neutral lipids in the postsynaptic membrane. Nakajima and Bridgman (1981) used the property of the antibiotic filipin to form large intramembranous filipin–sterol aggregates (De Kruijff and Demel, 1974), which can be readily detected by electron microscopy (e.g., Orci *et al.*, 1981) to study cholesterol localization in muscle membranes. The aggregates could be observed in the extrasynaptic regions of the sarcolemma, but were absent in receptor-rich areas leading the authors to conclude that the AChR environment was devoid of cholesterol. In contrast with Nakajima and Bridgman's study, morphological confirmation of the various biochemical studies demonstrating the presence of cholesterol in *Torpedo* receptor-rich membranes (Table I) has been produced by St. John *et al.* (1982). These authors indirectly established the presence of cholesterol in AChR membrane by showing the permeabilizing action of saponin. Saponin is known to interact with cholesterol-rich membrane regions; upon treatment with this reagent, molecules as large as ferritin could penetrate the otherwise sealed membrane vesicles. The contradicting reports summarized above point to the difficulties involved in this type of study. The lack of more specific probes contributes to the problem. Biophysical studies reviewed in the following section further emphasize the importance of cholesterol in AChR membranes.

The specificity of lipid–AChR interactions has also been addressed in lipid monolayer studies. In one case the penetration of purified AChR from *Torpedo* or *Electrophorus* into monolayers of pure lipids was used as an indicator of the degree of interaction (Popot *et al.*, 1978). AChR/monolayer interactions were found to be dominated by hydrophobic interactions; the nature of the polar moiety of the phospholipids had little influence on the rate or extent of AChR penetration. Incorporation into the monolayer was enhanced upon increasing the length of the fatty acyl chains of synthetic phosphatidylcholines (Popot *et al.*, 1978). Purified AChR showed a preference for cholesterol films as compared to monolayers prepared with any phospholipid or with several other types of sterols. Small differences in structure (such as those between cholesterol and ergosterol) have also been reported in the above study to influence AChR incorporation, indicating that the observed specificities are not merely the consequence of different bulk physical properties of the films. These findings show concurrence with two features of the lipids present in *Torpedo* membranes, i.e., the abundance of cholesterol and long-chain fatty acids (Table I and Fig. 2).

4.5. Phospholipid Chain Length, Saturation, and Head Group on AChR Affinity States and Receptor-Mediated Ion Flux

In an attempt to study in more detail the influence of lipid chemical structure on the affinity state transitions and the ion translocation function of the AChR, *Torpedo* receptor was reconstituted into liposomes of pure synthetic lipids (Criado *et al.*, 1982a; 1984). A critical concentration of 30%–40% of cholesteryl hemisuccinate was found to be necessary in the reconstituted system (liposomes made of this cholesteryl derivative and dimyristoyl PC) to mimic the kinetics of agonist-induced state transitions observed in native membranes. With increasing chain length of the saturated lecithins, a marked augmentation in carbamylcholine dissociation constants was observed. As shown in Fig. 4, the increase affected both the apparent equilibrium dissociation constants K_R and K_{eq}. Substitution by other dimyristoyl phospholipids for dimyristoyl PC had the same, though quantitatively less pronounced, effects (Fig. 5). Introduction of unsaturation in the acyl chains dramatically reverses the effect of increasing chain length (Fig. 6).

Another parameter measured, the $^{22}Na^+$ flux, provided information on the effect of lipid classes on the channel gating properties of the AChR. Unsaturated phosphatidylethanolamines in combination with 28–35 mol% of cholesteryl hemisuccinate was the best lipid mixture for reconstitution of the receptor-gating function in artificial lipid systems. When phosphatidylethanolamine was replaced totally or partially by other phospholipids with the same or different acyl chain composition, a marked decrease of ion transport was apparent, even when similar vesicle size, receptor incorporation, and agonist-induced affinity transitions were obtained. One of the conclusions to be drawn from this set of data is that the maintenance of the affinity state transitions of the reconstituted receptor is a necessary but not sufficient condition for the manifestation of the ion-gating receptor activity. A second conclusion is that the more unsatur-

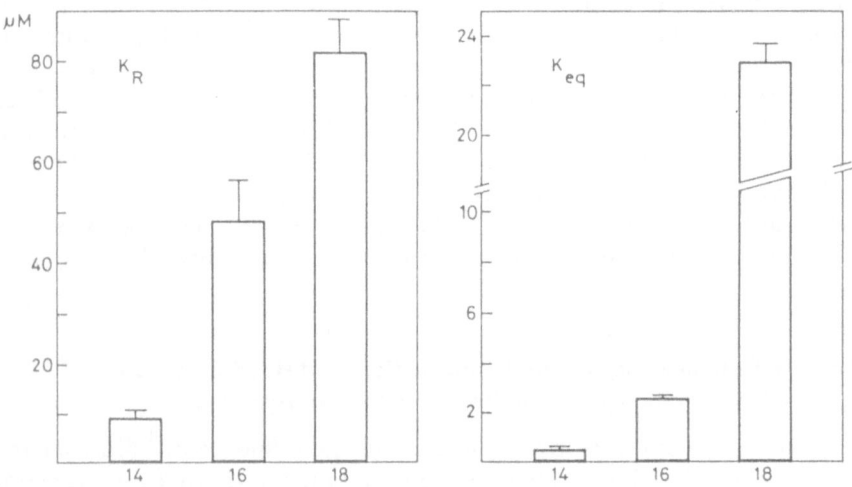

Figure 4. Increasing fatty acid chain length of phosphatidylcholines diminishes the affinity of the AChR for the agonist carbamylcholine. From Criado *et al.* (1984).

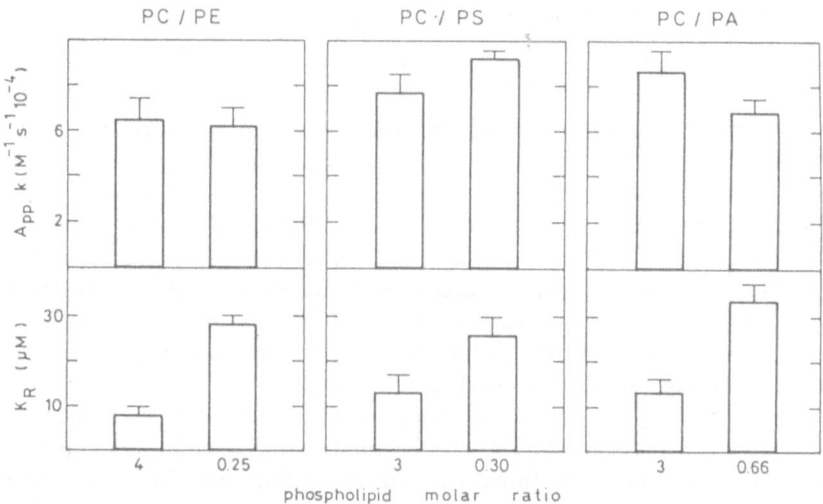

Figure 5. Apparent toxin association rates (k) are not significantly altered through substitution of dimyristoyl-phosphatidylcholine by other glycerophospholipids with different head group [ethanolamine (PE), serine (PS), and phosphatidic acid (PA)]. On the other hand, the apparent dissociation constant for the agonist carbamoylcholine at $t \to 0$ (K_R) tends to increase with increasing degree of substitution of choline by other head groups while maintaining the same fatty acid chain (14:0).

ated the acyl chains of phosphatidylethanolamine are, the higher the response that is observed, suggesting that a critical lipid packing is essential for the ion translocation function of the receptor. Thus the trends observed in our flux studies with synthetic lipids (Criado *et al.*, 1984) show concurrence with our more recent analysis of the lipid molecular species in native membranes (Rotstein *et al.*, 1987b) showing more than 75% polyenes in PE (of which the majority are hexaenoic molecular species, Fig. 2).

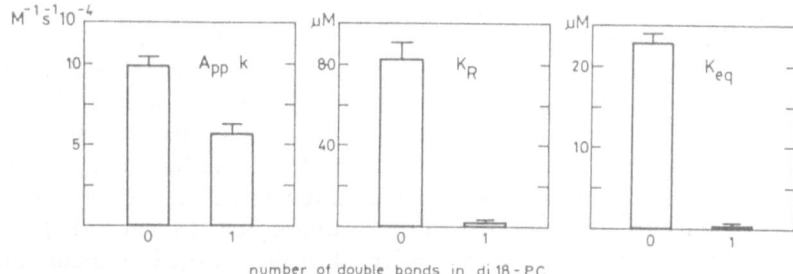

Figure 6. Introduction of a single double bond in dipalmitoyl phosphatidylcholine has a dramatic influence on the apparent dissociation constants of carbamoylcholine at $t \to 0$ (K_R) and $t \to \sim\infty$ (K_{eq}). The affinity of the agonist is much higher in a monosaturated lipid environment. Data from Criado *et al.* (1984).

5. DYNAMICS OF AChR AND LIPIDS IN THE MEMBRANE

5.1. Lipid Dynamics in AChR Membranes

In situ, the AChR is closely packed in the postsynaptic membrane, reaching densities of more than 10,000 particles of 8–9 nm diameter per μm^2 in the neuromuscular junction and even 50,000 units μm^2 in the electrocyte synapse (see review in Barrantes, 1979). AChR-rich membranes have a relatively high protein/lipid ratio (Table I), and packing habits appear to be conserved in AChR membranes, which are also densely covered by receptor particles. It is therefore not surprising to find clear indications of motionally restricted lipid in the electron spin resonance spectra of lipid spin labels incorporated into such membranes.

Early studies failed to detect immobilized lipid in the spectra of fatty acid derivatives incorporated into native membranes, leading to the postulation of a fluid lipid environment in the immediate vicinity of the AChR (Bienvenue *et al.*, 1977). Subsequently, ESR spectra of the stearic acid spin label 16-SASLX in AChR native membranes and in aqueous dispersions of the extracted membrane lipids enabled the observation of two components in differential ESR spectra, one of which corresponded to motionally restricted lipid in the membrane (Marsh and Barrantes, 1978). Similar immobilized components were observed with phospholipid (Marsh *et al.*, 1981) and steroid (Marsh and Barrantes, 1978; Ellena *et al.*, 1983) spin labels. Rousselet *et al.* (1979) had observed immobilization with fatty acid probes but not with phospholipid spin labels, from which they inferred specific binding of fatty acids to the AChR protein.

The work on reconstituted membranes by Ellena *et al.* (1983) finally settled this controversial issue. They demonstrated some degree of specificity in the association of the fatty acid and steroid labels with the receptor protein. Experiments with spin-labeled phosphatidylcholine, which shows no preferential selectivity in interaction with the protein as judged by ESR criteria clearly showed the immobilized component. Approximately 40 phospholipid molecules were motionally restricted by direct interaction with the hydrophobic surface of the protein (Table II). This result is in good agreement with estimates of the number of phospholipids which may be accommodated around the intramembranous perimeter of the AChR, based on structural data available to date. These data have been compiled on the basis of experiments undertaken in reconstituted systems, in which lipid–protein titrations can be readily accomplished (Marsh, 1985).

In the case of the AChR the projected surface of the intramembranous portion has recently been determined by electron microscopy at a resolution of about 2 nm (Brisson and Unwin, 1985). The corresponding radius of the membrane-embedded domain is about 8 nm. Thus the AChR nomomer could accommodate the motionally restricted lipid detected by ESR spec-

Table II. Stoichiometries of the Motionally Restricted Lipid Component in Reconstituted AChR Membranes and Other Lipid-Protein Systems[a]

Protein-membrane	M.W. $\times 10^{-3}$	N_1^{exp} (mol/mol)	$\dfrac{N_1^{exp}}{\sqrt{M.W.}}$	N_1^{calc} (mol/mol)	Ref.
Acetylcholine receptor-DOPC	250	40 ± 7	0.080 ±0.014	43	1
Na$^+$,K$^+$-ATPase-DOPC	314	63 ± 3	0.112 ±0.005	(~57-72)	2
Na$^+$,K$^+$-ATPase-shark rectal gland	265	58 ± 4	0.112 ±0.008	(~57-72)	3
Cytochrome oxidase-DMPC	165	45 ± 4	0.110 ±0.011	40-45	4
Ca^{2+}-ATPase-egg PC	115	22 ± 2	0.065 ±0.006	(~23-27)[b]	5
Sarcoplasmic reticulum-Ca^{2+}-ATPase	115	24	0.071	(~23-27)[b]	6
Bovine rod outer segment disk-rhodopsin	39	25 ± 3	0.125 ±0.016	24(±2)	7
Frog rod outer segment disk-rhodopsin	39	23 ± 2	0.114 ±0.010	(24)	3
Myelin proteolipid apoprotein-DMPC	25	10 ± 2	0.063 ±0.013	(~10-12)[c]	8

[a] Modified from data compiled by Marsh (1985).
[b] Calculated assuming a dimer, with monomer radius 20 Å (Ref. 9).
[c] Calculated assuming hexamer; references: 1, Ellena et al. (1983); 2, Brotherus et al. (1981); 3, Marsh et al. (1982); 4, Knowles et al. (1979); 5, Silvius et al. (1984); 6, Thomas et al. (1982); 7, Watts et al. (1979); 8, Brophy et al. (1984); 9, Marsh and Watts (1982). Abbreviations used; DOPC, dioleoyl-phosphatidylcholine; DMPC, dimyristoyl phosphatidylcholine.

troscopy within the first shell around its perimeter. The calculated number of lipid molecules that can be accommodated around the intramembranous perimeter of the AChR and other intramembranous proteins is also given in Table II.

It is now generally accepted that the strong immobilization of the AChR in the postsynaptic membrane arises from protein-protein rather than protein-lipid interactions (Cartaud et al., 1981; reviewed by Barrantes, 1983). The immobilization of the protein obviously has consequences on lipid mobility: the rotational correlation times of the motionally restricted lipid in native AChR membranes are about one-tenth that of the fluid lipids (Marsh and Barrantes, 1978). Both lipid components, however, are relatively more mobile than an average integral membrane protein (~20 μs). For a correlation time of about 50 nsec (Marsh and Barrantes, 1978) the lipid-lipid exchange rate in a fluid bilayer is estimated to be $\sim 2 \times 10^7 \sec^{-1}$. The rotational correlation time compounds motional contributions of (1) the lipid acyl chains interacting directly with the protein and (2) the exchange of lipid on and off the protein surface. The effective lifetime of the lipids on the protein boundary has a lower limit of about 50 nsec shorter than the lifetime of average lipid–lipid exchange in fluid lipid bilayers resulting from *lateral* diffusion. Thus lipids in the immediate vicinity of the AChR exhibit motion restricted in comparison to that of the fluid phase, and *independent* of the exchange by diffusion (see below).

5.2. Lateral Diffusion of AChR and Lipids in Natural and Reconstituted Systems

The so-called "free area theories" envisage diffusion of a lipid molecule as the resultant of two combined steps: (1) formation of a free volume (a sort of hole) adjacent to the lipid molecule in question, and (2) subsequent filling of the free volume created by the diffusant molecule jumping into the hole. Free area theories have been used to explain the lateral diffusion of fluorescent lipid analogs in reconstituted AChR liposomes and in corresponding bilayers devoid of the receptor protein. NBD-phosphatidylethanolamine exhibited values of D, the lateral diffusion coefficient of 8×10^{-8} cm^2 sec^{-1} (Criado *et al.*, 1982b). In the case of the AChR reconstituted into artificial bilayers at relatively low protein/lipid ratios, the lateral diffusion of the protein can be estimated by the Saffman and Delbrück (1975) formula. This is based on the Stokes–Einstein treatment of motion in a solvent medium, in which lateral motion can be envisaged as originating in random collisions of the macromolecule with the solvent molecules, and is opposed by frictional forces with the same. The diffusion coefficient, D, is given by $D = kT/f$, where k is the Boltzman constant and f is the frictional coefficient. In the case of a spherical molecule of radius r diffusing in a medium of viscosity, $D = kT/6\pi\eta r$ (the Stokes–Einstein equation). Saffman and Delbrück applied this formula to the case of a cylinder of radius r oriented perpendicular to and diffusing in a thin viscous sheet of viscosity η and thickness h (equal to the height of the embedded cylinder), a treatment that is more relevant to a protein traversing a membrane, and the AChR in particular. The diffusion coefficient becomes

$$D = (kT/4\pi\eta_m h) \ln(\eta_m h / \eta_w r - 0.5772)$$

where η_m is the viscosity of the viscous sheet ("the membrane") and η_w ($\sim 10^{-2}$ poise) is the viscosity of the aqueous medium in which the sheet is immersed. The sheet is treated as a continuum of solvent molecules. The formula predicts a weak dependence of D on protein radius, a prediction that appears to be followed roughly by the few integral membrane proteins studied to date, including the AChR (Criado et al., 1982b).

It should be noted that the theory finds validity only in the case of very diluted proteins (as with proteins reconstituted in lipid bilayers); the rate of diffusion of the AChR monomers and dimers lies in the range of $1-3 \times 10^{-8} \, cm^2 \, sec^{-1}$ in lipid bilayers in the liquid-crystalline phase and within the 14–37°C range (Criado et al., 1982b). The presence of cholesterol does not appear to modify D.

In the plasmalemma of developing rat myotubes the situation is somewhat different. Two populations of AChR molecules are observed: a diffusely distributed population having $D = 5 \times 10^{-11} \, cm^2 \, sec^{-1}$, and a localized, patched population with $D \sim 10^{-12} \, cm^2 \, sec^{-1}$ (Axelrod et al., 1976). In blebs of myoblast sarcolemma, Tank et al. (1982) found D to be $3 \times 10^{-9} \, cm^2 \, sec^{-1}$, whereas Poo (1982) estimated D to be $2.6 \times 10^{-9} \, cm^2 \, sec^{-1}$ in Xenopus embryonic myoblasts. D in reconstituted membranes is at least 50-fold higher than the highest values found in native membranes. That is, the AChR in synthetic lipid membranes undergoes essentially free lateral diffusion. It does not exhibit a large decrease in lateral mobility as compared to the lipids. Reconstituted membranes differ from natural membranes in their lipid composition, concentration of protein, and absence of peripheral proteins. It is likely that the extensive protein–protein interactions occurring both between AChR molecules on the one hand and between AChR molecules and other nonreceptor proteins on the other are mainly responsible for the relative immobilization of AChR in the native membrane and particularly in the synaptic region.

Considerations similar to those outlined above for the translation mobility of the AChR are also applicable to the rotational motion of the protein. This has been studied by means of electron spin resonance and phosphorescence anisotropy approaches; the rotational correlation time ϕ of the monomeric AChR is about 10–25 μsec (Bartholdi et al., 1981), in full agreement with the expected value. Slower relaxation times are probably related to the existence of higher oligomeric AChR species. Since the translational motion of the AChR in reconstituted lipid bilayers can be practically considered a free, unrestricted lateral diffusion, D can be related to the rotational relaxation time by $D \sim 4r^2/\phi$.

REFERENCES

Anderson, D. J., Tzartos, S. J., Gullick, W., Lindstrøm, J., and Blobel, G., 1983, J. Neurosci. 3:1773–1784.

Axelrod, D., Koppel, D. E., Schlessinger, J., Elson, E., and Webb, W. W., 1976, Biophys. J. 16:1055–1069.

Ballivet, M., Patrick, J., Lee, J., and Heinemann, S., 1982, *Proc. Natl. Acad. Sci. USA* **79**:4466-4470.

Barrantes, F. J., 1979, *Ann. Rev. Biophys. Bioeng.* **8**:287-321.

Barrantes, F. J., 1982a, in: *Neuroreceptors* (F. Hucho, ed.), pp. 315-328, W. de Gruyter, Berlin.

Barrantes, F. J., 1982b, *J. Cell Biol.* **92**:60-68.

Barrantes, F. J., 1983, *Int. Rev. Neurobiol.* **24**:259-341.

Barrantes, F. J., 1986, in: *Ionic Channels in Cells and Model Systems* (R. Latorre, ed.), Plenum Press, New York.

Barrantes, F. J., Neugebauer, D.-Ch., and Zingsheim, H. P., 1980, *FEBS Lett.* **112**:3-78.

Bartholdi, M., Barrantes, F. J., and Jovin, T. M., 1981. *Eur. J. Biochem.* **120**:389-397.

Bell, R. M., and Coleman, R. A., 1980, *Ann. Rev. Biochem.* **43**:243-277.

Bienvenue, A., Rousselet, A., Kato, G., and Devaux, P. F., 1977, *Biochemistry* **16**:841-848.

Bleasdale, J. E., Hawthorne, J. N., Widlund, L., and Heilbronn, E., 1976, *Biochem. J.* **158**:557-565.

Boheim, G., Hanke, W., Barrantes, F. J., Eibl, H., Sakmann, B., Fels, G., and Maelicke, A. 1981, *Proc. Natl. Acad. Sci. USA* **78**:3586-3590.

Bon, F., Lebrun, E., Gomel, J., van Rapenbusch, R., Cartaud, J., Popot, J.-L., and Changeux, J.-P., 1984, *J. Mol. Biol.* **176**:205-237.

Breckenridge, W. C., and Vincendon, G., 1971, *C. R. Acad. Sci. Paris* **273D**:1337-1339.

Brisson, A., 1980, These Doctorat d'Etat, Grenoble, France.

Brisson, A., and Unwin, P. N. T., 1985, *Nature* **315**:474-477.

Brophy, P. J., Horváth, L. I., and Marsh, D., 1984, *Biochemistry* **23**:860-865.

Brotherus, J. R., Griffith, O. H., Brotherus, M. O., Jost, P. C., Silvius, J. R., and Hoking, L. E., 1981, *Biochemistry* **20**:5261-5267.

Cartaud, J., Sobel, A., Rousselet, A., Devaux, P. F., and Changeux, P. F., 1981, *J. Cell Biol.* **90**:418-426.

Changeux, J.-P., Bon, F., Cartaud, J., Devillers-Thiery, A. J., Heidmann, T., Holton, B., Ngheim, H.-O., Popot, J.-L., van Rapenbusch, R., and Tzartos, S., 1983, *Cold Spring Harbor Symp. Quant. Biol.* **68**:35-52.

Claudio, T., Ballivet, M., Patrick, J., and Heinemann, S., 1983, *Proc. Natl. Acad. Sci. USA* **80**:1111-1115.

Conti-Tronconi, B. M., Hunkapiller, M. W., and Raftery, M. A., 1984, *Proc. Natl. Acad. Sci. USA* **81**:2631-2634.

Conti-Tronconi, B. M., and Raftery, M. A., 1982, *Ann. Rev. Biochem.* **51**:491-530.

Criado, M., Eibl, H., and Barrantes, F. J., 1982a, *Biochemistry* **21**:3622-3629.

Criado, M., Eibl, H., and Barrantes, F. J., 1984, *J. Biol. Chem.* **259**:9188-9198.

Criado, M., Hochswender, S., Sarin, V., Fox, J. L., and Lindstrøm, J., 1985, *Proc. Natl. Acad. Sci. USA* **82**:2004-2008.

Criado, M., Vaz, W. L. C., Barrantes, F. J., and Jovin, T. M., 1982b, *Biochemistry* **21**:5750-5755.

Dalziel, A. W., Rollins, E. S., and McNamee, M. G., 1980, *FEBS Lett.* **122**:193-196.

De Kruijff, B., and Demel, R. A., 1974, *Biochim. Biophys. Acta* **339**:57-70.

Devillers-Thiery, A., Giraudat, J., Bentavoulet, M., and Changeux, J-P., 1983, *Proc. Natl. Acad. Sci. USA* **80**:2067-2071.

Doster, W., Hess, B., Watters, D., and Maelicke, A., 1980, *FEBS Lett.* **113**:312-314.

Ellens, J. F., Blazing, M. A., and McNamee, M. G., 1983, *Biochemistry* **22**:5523-5535.

Fairclough, R. H., Finer-Moore, J., Love, R. A., Kristofferson, D., Desmeules, P. J., and Stroud, R. M., 1983, *Cold Spring Harbor Symp. Quant. Biol.* **68**:9-20.

Finer-Moore, J., and Stroud, R. M., 1984, *Proc. Natl. Acad. Sci. USA* **81**:155-159.

Froehner, S. C., 1981, *Biochemistry* **20**:4905-4915.

Giraudat, J., Montecucco, C., Bisson, R., and Changeux, J.-P., 1985, *Biochemistry* **24**:3121-3127.

González-Ros, J. M., Llanillo, M., Paraschos, A., and Martinez-Carrion, M., 1982, *Biochemistry* **21**:3467-3474.

Guy, H. R., 1981, *Cell Molec. Neurobiol.* 1:231-258.

Guy, H. R., 1983, *Biophys. J.* 45:249-261.

Hamill, O. P., Marty, A., Neher, E., Sakmann, B., and Sigworth, F. J., 1981, *Pflüggers Arch. ges. Physiol.* 391:85-100.

Hamilton, S. L., Pratt, D. R., and Eaton, D. C., 1985, *Biochemistry* 24:2210-2219.

Heuser, J. E., and Salpeter, S. R., 1979, *J. Cell Biol.* 82:150-173.

Holtzman, E., Wise, D., Wall, J., and Karlin, A., 1982, *Proc. Natl. Acad. Sci. USA* 79:310-314.

Hopp, T. P., and Woods, K. R., 1981, *Proc. Natl. Acad. Sci. USA* 78:3824-3828.

Juillerat, M., Barkas, T., and Tzartos, S., 1984, *FEBS Lett.* 168:143-148.

Kao, P. N., Dwork, A. J., Kaldany, R.-R. J., Silver, M. L., Wideman, J., Stein, S., and Karlin, A., 1984, *J. Biol. Chem.* 259:11662-11665.

Karlin, A., Holtzman, E., Yodh, N., Lobel, P., Wall, J., and Hainfield, J., 1983, *J. Biol. Chem.* 258:6678-6681.

Kilian, P. L., Dunlap, C. R., Mueller, P., Schell, M. A., Huganir, R. L., and Racker, E., 1980, *Biochem. Biophys. Res. Commun.* 93:409-414.

Kistler, J., Stroud, R. M., Klymkowsky, M. W., Lalancette, R. A., and Fairclough, R. H., 1982, *Biophys. J.* 37:371-383.

Klymkowsky, M. W., and Stroud, R. M., 1979, *J. Mol. Biol.* 128:319-334.

Knowles, P. F., Watts, A., and Marsh, D., 1979, *Biochemistry* 18:4490-4487.

Kosower, E. M., 1983, *FEBS Lett.* 155:245-247.

Kosower, E. M., 1984, *FEBS Lett.* 172:1-5.

Kostic, D., Rakic, L., and Vranesevic, A., 1972, *Evol. Biokhim. Fisiol.* 8:494-498.

Kubo, T., Nods, M., Takai, T., Tanabe, T., Kayano, T., Shimizu, S., Tanaka, K., Takahashi, H., Hirose, T., Inayama, S., Kikuno, R., Mivara, T., and Numa, S., 1985, *Eur. J. Biochem.* 149:5-13.

Kyte, J., and Doolittle, R. F., 1982, *J. Mol. Biol.* 157:105-132.

Lantz, G., Burgun, C., Cremel, G., Hubert, P., Darcy, F., and Waksman, A. 1985, *Neurochem. Int.* 7:331-339.

LaRochelle, W. J., Wray, B. E., Sealock, R., and Froehner, S. C., 1985, *J. Cell Biol.* 100:684-691.

Latorre, R., Alvarez, O., Cecchi, X., and Vergara, C., 1985, *Ann. Rev. Biophys. Chem.* 14:79-111.

Lennon, V. A., McCormick, D. J., Lambert, E. H., Griesmann, G. E., and Atassi, M. Z., 1985, *Proc. Natl. Acad. Sci. USA* 82:8805-8809.

Letters, R. 1964, *Biochem. J.* 93:313-316.

Lindstrøm, J., Criado, M., Hochschwender, S., Fox, J. L., and Sarin, V., 1984, *Nature* 311:573-575.

Lindstrøm, J., Tzartos, S., and Gullick, W., 1982, *Ann. N.Y. Acad. Sci.* 377:1-19.

Lindstrøm, J., Tzartos, S., Gullick, W., Hochschwender, S., Swanson, L., Sargent, P., Jacob, M., and Montal, M., 1983, *Cold Spring Harbor Symp. Quant. Biol.* 68:89-99.

Lo, M. M. S., Barnard, E. A., and Dolly, J. O., 1982, *Biochemistry* 21:2210-2217.

Lo, M. M. S., Garland, P. B., Lamprecht, J., and Barnard, E. A., 1980, *FEBS Lett.* 111:407-412.

Lüdi, H., Oeticker, H., Brodbeck, U., Ott, P., Schwendimann, B., and Fulpius, B. W., 1983, *J. Membrane Biol.* 74:75-84.

Lynch, D. V., and Thompson, G. A., Jr., 1984, *Trends Biochem. Sci.* 9:442-445.

Marsh, D., 1985, in: *Progress in Protein-Lipid Interactions* (A. Watts and J. J. H. H. M. De Pont, eds.), Vol. 4, pp. 143-172, Elsevier Science, Amsterdam.

Marsh, D., and Barrantes, F. J., 1978, *Proc. Natl. Acad. Sci. USA* 75:4329-4344.

Marsh, D., and Watts, A., 1982, in: *Lipid-Protein Interactions* (P. C. Jost and O. H. Griffith, eds.), Vol. II, pp. 53-126, Wiley, New York.

Marsh, D., Watts, A., and Barrantes, F. J., 1981, *Biochim. Biophys. Acta* 645:97-101.

Marsh, D., Watts, A., Pates, R. D., Uhl, R., Knowles, P. F., and Esmann, M., 1982, *Biophys. J.* 37:265-274.

McCormick, D. J., and Atassi, M. Z., 1984, *Biochem. J.* 224:995-1000.

McNamee, M. G., Ellena, J. F., and Dalziel, A. W., 1982, *Biophys. J.* **37**:103-104.
Michaelson, D. M., and Raftery, M. A., 1974, *Proc. Natl. Acad. Sci. USA* **71**:4768-4772.
Middlemas, D. S., and Raftery, M. A., 1983, *Biochim. Biophys. Res. Commun.* **115**:1075-1082.
Mishina, M., Tobimatsu, T., Imoto, K., Tanaka, K., Fujita, Y., Fukuda, K., Kurasaki, M., Takahashi, H., Morimoto, Y., Hirose, T., Inayama, S., Takahashi, T., Kuno, M., and Numa, S., 1985, *Nature* **318**:364-368.
Momoi, M. Y., and Lennon, V. A., 1986, *J. Neurochem.* **46**:76-81.
Nakajima, Y., and Bridgman, P. C., 1981, *J. Cell Biol.*, **88**:453-458.
Naner, E., and Sakmann, B., 1976, *Nature* **260**:799-802.
Nghiêm, H. O., Cartaud, J., Dubrenil, C., Kordeli, C., Buttin, G., and Changeux, J.-P., 1983, *Proc. Natl. Acad. Sci. USA* **80**:6403-6407.
Noda, M., Furutani, Y., Takahashi, H., Toyosato, M., Tanabe, T., Shimizu, S., Kikyotani, S., Kayano, T., Hirose, T., Inayama, S., and Numa, S., 1983a, *Nature* **305**:818-823.
Noda, M., Takahashi, H., Tanabe, T., Toyosato, M., Furutani, Y., Hirose, T., Asai, M., Inayama, S., Miyata, T., and Numa, S., 1982, *Nature* **299**:793-797.
Noda, M., Takahashi, H., Tanabe, T., Toyosato, M., Kikyotani, S., Hirose, T., Asai, M., Takashima, H., Inayama, S., Miyata, T., and Numa, S., 1983b, *Nature* **301**:251-255.
Noda, M., Takahashi, H., Tanabe, T., Toyosato, M., Kikyotani, S., Furutani, Y., Hirose, T., Takashima, H., Inayama, S., Miyata, T., and Numa, S., 1983c, *Nature* **302**:528-532.
Numa, S., Noda, M., Takahashi, H., Tanabe, T., Toyosato, M., Furutani, Y., Kikyotani, S., 1983, *Cold Spring Harbor Symp. Quant. Biol.* **48**:57-69.
Orci, L., Montesano, R., Meda, P., Malaisse-Lagae, F., Brown, D., Perrelet, A., and Vassalli, P., 1981, *Proc. Natl. Acad. Sci. USA* **78**:293-297.
Poo, M., 1982, *Nature* **295**:332-334.
Popot, J. L., Demel, R. A., Sobel, A., Van Deenen, L. L. M., and Changeux, J. P., 1978, *Eur. J. Biochem.* **85**:27-42.
Popot, J. L., and Changeux, J.-P., 1984, *Physiol. Rev.* **64**:1162-1239.
Raftery, M. A., Hunkapiller, M. W., Strader, C. D., and Hood, L. E., 1980, *Science* **208**:1454-1457.
Reynolds, J. A., and Karlin, A., 1978, *Biochemistry* **17**:2035-2038.
Rosenberg, P., 1973, *J. Pharmaceut. Sci.* **62**:1552-1554.
Rotstein, N., Arias, H., Aveldaño, M. I., and Barrantes, F. J., 1987a, *J. Neurochem.* **49**:1341-1347.
Rotstein, N., Arias, H., Barrantes, F. J., and Aveldaño, M. I., 1987b, *J. Neurochem.* **49**:1333-1340.
Rousselet, A., Cartaud, J., and Devaux, P. F., 1981, *Biochim. Biophys. Acta* **648**:169-185.
Rousselet, A., Devaux, P. F., and Wirtz, K. W., 1979, *Biochem. Biophys. Res. Commun.* **90**:871-877.
Sapatini, D. D., Kreivich, G., Morimoto, T., and Adesnik, M., 1982, *J. Cell Biol.* **92**:1-22.
Saffman, P. G., and Delbrück, M., 1975, *Proc. Natl. Acad. Sci. USA* **72**:3111-3113.
Sakmand, B., Methfessel, C., Mishina, M., Takahashi, T., Takai, T., Kurasaki, M., Fukuda, K., and Numa, S., 1985, *Nature* **318**:538-543.
Sator, U. S., Gonzàlez-Ros, J. M., Calvo-Fernandez, P., and Martinez-Carrion, M., 1979, *Biochemistry* **18**:1200-1206.
Schiebler, W., and Hucho, F., 1978, *Eur. J. Biochem.* **85**:55-63.
Schindler, H., 1982, *Neurosc. Res. Program Bull.* **20**:295-301.
Sealock, R., Wray, B. E., and Froehner, S. C., 1984, *J. Cell Biol.* **96**:2239-2244.
Shibahara, S., Kubo, T., Perski, H. J., Takahashi, H., Noda, M., and Numa, S., 1985, *Eur. J. Biochem.* **146**:15-22.
Silvius, J. R., McMillen, D. A., Saley, N. D., Jost, P. C., and Griffith, O. H., 1984, *Biochemistry* **23**:538-547.

St. John, P. A., Froehner, S. C., Goodenough, D. A., and Cohen, J. B., 1982, *J. Cell Biol.* **92:**333-342.

Strader, C. D., and Raftery, M. A., 1980, *Proc. Natl. Acad. Sci. USA* **77:**5807-5811.

Sumikawa, K., Houghton, M., Smith, J. C., Bell, L., Richards, B. M., and Barnard, E. A., 1982, *Nucleic Acids Res.* **10:**5809-5822.

Takai, T., Noda, M., Furutani, Y., Takahashi, H., Notake, M., Shimizu, S., Kayano, T., Tanabe, T., Tanaka, K., Hirose, T., Inayama, S., and Numa, S., 1984, *Eur. J. Biochem.* **143:**109-115.

Takai, T., Noda, M., Mishina, M., Shimizu, S., Furutani, Y., Kayano, T., Ikeda, T., Kubo, T., Takahashi, T., Kuno, M., and Numa, S., 1985, *Nature* **315:**761-764.

Tanabe, T., Noda, M., Furutani, Y., Takai, T., Takahashi, H., Tanaka, K., Hirose, T., Inayama, S., and Numa, S., 1984, *Eur. J. Biochem.* **144:**11-17.

Tank, D. W., Wu, E.-S., and Webb, W. W., 1982, *J. Cell Biol.* **92:**207-212.

Tarrab-Hazdai, R., Bercovici, T., Goldfarb, V., and Gitler, C., 1980, *J. Biol. Chem.* **255:**1204-1209.

Thomas, D. D., Bigelow, D. J., Squier, T. C., and Hidalgo, C., 1982, *Biophys. J.* **37:**217-225.

Trams, E. G., and Brown, E. A. B., 1972, *Jugoslav. Physiol. Pharmacol. Acta* **8:**97-105.

Verstraete, M., and Vandenbroucke, J., 1956, *Acta Med. Scand.* **155:**37-48.

Watts, A., Volotovski, I. D., and Marsh, D., 1979, *Biochemistry* **18:**5006-5013.

White, M. M., Mixter Mayne, K., Lester, H., and Davidson, N., 1985, *Proc. Natl. Acad. Sci. USA* **82:**4852-4856.

Wilson, P. T., Lentz, T. L., and Hawroth, E., 1985, *Proc. Natl. Acad. Sci. USA* **82:**8790-8794.

Wise, D. S., Wall, J., and Karlin, A., 1981, *J. Biol. Chem.* **256:**12624-12627.

Young, E. F., Ralston, E., Blake, J., Ramachandran, J., Hall, Z. W., and Stroud, R. M., 1985, *Proc. Natl. Acad. Sci. USA* **82:**626-630.

Zingsheim, H. P., Neugebauer, D.-Ch., Barrantes, F. J., and Frank, J., 1980, *Proc. Natl. Acad. Sci. USA* **77:**952-956.

Zingsheim, H. P., Neugebauer, D.-Ch., Frank, J., Hänicke, W., and Barrantes, F. J., 1982a, *EMBO J.* **1:**541-547.

Zingsheim, H. P., Barrantes, F. J., Frank, J., Hänicke, W., and Neugebauer, D.-Ch., 1982b, *Nature* **299:**81-84.

Chapter 6

Fluorescence Spectroscopy in the Study of Sarcoplasmic Reticulum Calcium-ATPase

Sergio Verjovski-Almeida and Eleonora Kurtenbach

1. INTRODUCTION

Sarcoplasmic reticulum is an intracellular membrane system that serves as a model for active transport of Ca^{2+} and coupled enzyme catalysis in biological membranes. The isolation of vesicular fragments of sarcoplasmic reticulum membranes from muscle homogenates and the demonstration that they were capable of maintaining calcium transport at the expense of the hydrolysis of ATP (Ebashi and Lipman, 1962; Hasselbach and Makinose, 1961) has led the way to extensive characterization of the kinetics and mechanism of calcium transport and energy transduction in this system. The work with sarcoplasmic reticulum has been extensively reviewed (de Meis, 1980; Ikemoto, 1982; Inesi, 1985). The sarcoplasmic reticulum membranes have a highly specific protein composition, consisting of a Ca^{2+}-dependent ATPase of 115,000 daltons per polypeptide chain. The ATPase accounts for 80%–90% of the total protein, and its localization within the membrane has been characterized both by X-ray diffraction from oriented multilayers (Dupont *et al.*, 1973; Herbette *et al.*, 1977) and by electron microscopy (Deamer and Baskin, 1969; Hasselbach and Elfvin, 1967). These

SERGIO VERJOVSKI-ALMEIDA AND ELEONORA KURTENBACH • Departamento de Bioquímica, Instituto de Ciencias Biomédicas, Universidade Federal do Rio de Janeiro, 21910 Rio de Janeiro, Brazil.

studies have demonstrated that the ATPase is densely spaced within the plane of the membrane and that the ATPase chain has an amphiphylic character, with a hydrophobic portion inserted into the membrane bilayer and a hydrophylic portion protruding from the cytoplasmic surface of the membrane into the aqueous phase. More recently, the amino acid sequence of the ATPase has been established by MacLennan *et al.* (1985) and a computer-simulated structural model has been proposed from the primary sequence.

Fluorescence spectroscopy is a sensitive method that allows the detection of conformational changes and interactions of macromolecules, and it has been used in the study of sarcoplasmic reticulum ATPase. A number of interesting structural features of this enzyme have been characterized by fluorescence, some of which are given here as an example of the biochemical applications of fluorescence spectroscopy.

2. STRUCTURE AND CONFORMATION

2.1. Intrinsic Fluorescence of Aromatic Amino Acid Residues

In this section we will focus our attention on the origins of fluorescence emission by proteins and the main causes of its variation. Extensive data have been published on the fluorescence of numerous proteins and peptides. We will discuss the kind of structural and conformational information that can be obtained by fluorescence spectroscopy studies using the sarcoplasmic reticulum Ca^{2+}-ATPase as a model.

Natural protein fluorescence is a result of the presence of intrinsic fluorophores in this macromolecule, these being tyrosine, tryptophan, and phenylalanine. The tryptophan residues of proteins generally account for about 90% of the total fluorescence emitted and are highly sensitive to solvent polarity being subject to both general and specific solvent effects. As a result, the emission spectra of tryptophan residues can reflect the polarity of their surrounding environment. Thus, the emission spectra of proteins are sensitive to the binding of substrates, association reaction, and denaturation, for example.

If protein is excited by uv light its absorption spectrum shows a maximum near 280 nm that corresponds to both tyrosine and tryptophan residues. Excitation of individual amino acids at 280 nm presents a fluorescence emission spectrum that is shown in Fig. 1. As we can see, the emission of tyrosine in water occurs at 303 nm and it is described as relatively insensitive to solvent polarity. The emission maximum of tryptophan in water occurs at 348 nm, and is highly dependent upon polarity (Teale and Weber, 1957).

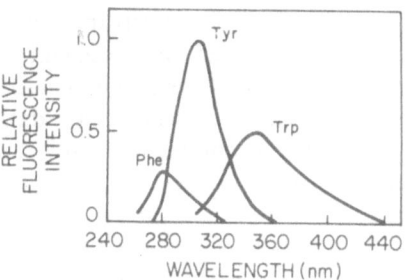

Figure 1. Fluorescence emission spectra of the aromatic amino acids. The emission maximum of phenylalanine in water occurs at 282 nm, of tyrosine at 303 nm, and the emission maximum of tryptophan in water is at 348 nm. (From Teale and Weber, 1957.)

In 1976 Dupont showed, for the first time, that sarcoplasmic reticulum vesicles exhibit a high fluorescence yield presenting an emission fluorescence spectrum with an optimum at 335 nm characteristic of tryptophanyl residues. Shown in Fig. 2 is a good resolution spectrum of excitation and emission by either sarcoplasmic reticulum or purified Ca^{2+}-ATPase (Verjovski-Almeida, 1981). It is known from amino acid content determinations that the Ca^{2+}-ATPase has about 35 hydrophobic tyrosine and tryptophan aromatic side-chain amino acid residues per polypeptide chain. We can see

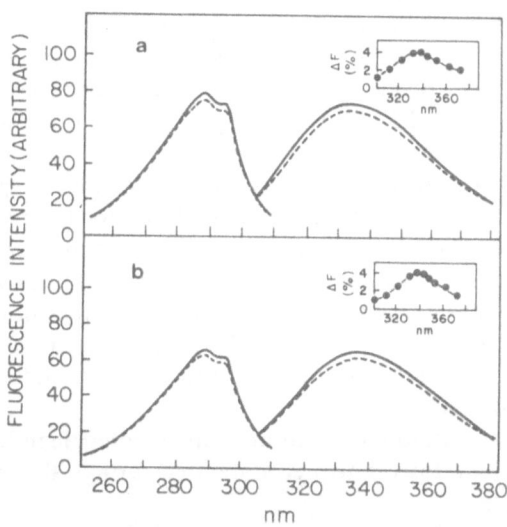

Figure 2. Excitation and emission spectra of sarcoplasmic reticulum ATPase. Vesicles (a) or purified ATPase (b) were suspended at a protein concentration of 100 μg/ml in a medium containing 100 mM KCl, 20 mM MOPS buffer, pH 7.0, 5 mM $MgCl_2$, and 0.5 mM EGTA (solid lines). After addition of 1.0 mM $CaCl_2$ (dashed lines), spectral measurements were repeated. In the excitation spectra emission was measured at 335 nm. In the emission spectra excitation was at 288 nm. The insets show the differences between the two emission spectra (ΔF) obtained in the absence and presence of calcium. (From Verjovski-Almeida, 1981.)

that the excitation spectrum exhibited two optima at 288 and 295 nm and an optimum at 335 nm in the emission fluorescence spectrum corresponding to the peaks of tryptophanyl residues of the protein. In order to evaluate the significance of fluorescence data it is important to consider the properties of fluorescent molecules and the theoretical treatment of fluorescence spectroscopy.

Quantum mechanics tells us that energy of the electrons of a molecule can only be of certain values corresponding to certain distributions of the electrons about the nuclei. Each of these distributions corresponds to an electronic state of the molecule. At room temperature almost all molecules in equilibrium with the surroundings are found in the electronic state of the lowest energy, i.e., the ground state (S_0). States of higher energy are referred to as excited electronic states $(S_1, S_2, \ldots, T_1)$, where S is denominated singlet excited state and T is the triplet excited state.

Promotion of a molecule from its ground electronic state to one or another excited electronic state occurs upon absorption of a photon by the ground state. An excited state molecule may absorb a photon and be promoted to a still higher level. It is interesting to remember that for a photon to be absorbed there must be an interaction between the molecule and the light and that only photons of energies corresponding to the energy difference, ΔE, between the lower and upper electronic states of the molecule can be absorbed, i.e., this process is said to be quantificated. For each electronic state of a molecule there exists a set of vibrational modes that is associated with a set of energy levels. As a result the absorption spectra appear as multiple peaks corresponding to vibrational modes associated with an electronic state (see Fig. 3).

As we know the electrons occupy specific regions of space around the nuclei that are called orbitals. Each orbital cannot be occupied by more than two electrons and these must have opposite spins (paired electrons). When a molecule is excited by a photon one electron from a ground orbital "jumps" to a higher-energy orbital. If this electron does not change its spin orientation the final excited state is called singlet (paired electron spins). If this electron changes its spin orientation the final state is called triplet (unpaired electron spins).

Spontaneous emission of a photon from a molecule in its lowest excited singlet state (S_1) or lowest triplet state (T_1) brings the molecule to the ground state (S_0).

The well-known fluorescence emissions of the organic molecules correspond to the allowed electronic transition $S_1 \rightarrow S_0$, and the return to the ground state from an excited state does not require an electron to change its spin orientation. The lifetime of the excited state for an allowed transition like $S_1 \rightarrow S_0$ may be expected to be in the range of nanoseconds (1 nsec $= 10^{-9}$ sec). Most of the lifetimes experimentally determined fall in the region

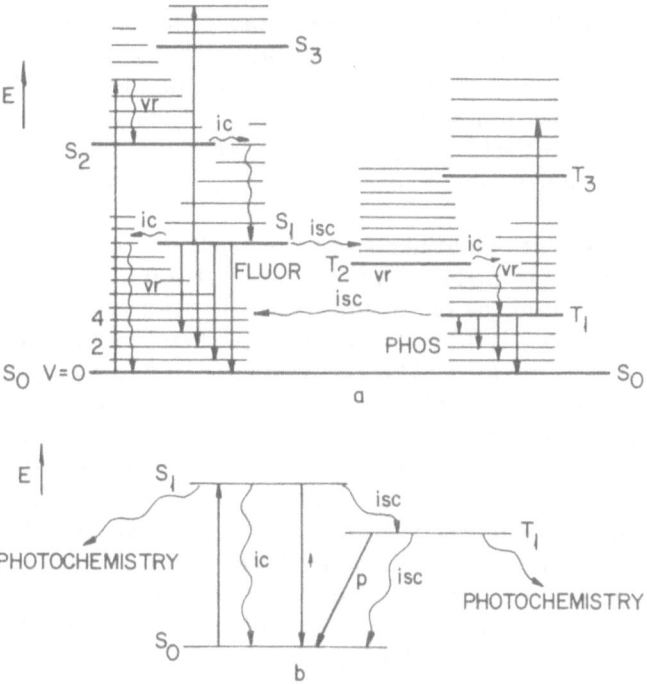

Figure 3. The energy-level diagram for an organic molecule. The ground, first, second, and third electronic singlet states are depicted by S_0, S_1, S_2, and S_3, respectively (heavy horizontal lines). For each of these electronic energy levels there exists a number of vibrational levels (light horizontal lines). Radiative transitions are shown as straight downward arrows. Radiationless relaxation processes are shown as wavy arrows. vr, vibrational relaxation; ic, internal conversion; isc, intersystem crossing. Fluorescence is indicated as f or FLUOR and phosphorescence as p or PHOS. In the upper scheme the molecule is excited to S_2 by absorption of a photon. Excited molecules rapidly relax to the lowest excited state S_1, by internal conversion (10^{-12} sec) and vibration relaxation. Molecules in S_1 either fluoresce (10^{-7}–10^{-9}), undergo intersystem crossing in this case to T_2, or convert radiationlessly to S_0. Molecules in the T_2 state can relax to the lowest triplet state T_1 by internal conversion and from there they can emit phosphorescence (10^{-3}–10^2 sec). In the upper scheme we also show that an excited state molecule may absorb a photon and be promoted to a still higher level as $S_1 \rightarrow S_3$ and $T_1 \rightarrow T_3$. In the lower scheme a simplified relaxation path shows the most relevant processes that govern the relaxation of an electronically excited molecule.

of 1–20 nsec. The factors that modify the fluorescence emission must act within this time interval in order to be at all effective.

Molecules in the S_1 state can also undergo conversion to the first triplet state T_1. Emission from T_1 to S_0 is called phosphorescence and a change in spin orientation is necessarily needed for this. Generally this emissive process is shifted to longer wavelengths (lower energy) if compared to the fluorescence. Transition from T_1 to the ground state occurs with low probability, and as a result, the rate constant for such emission is several orders

of magnitude smaller than those of fluorescence. Thus, the phosphorescence
lifetimes range from milliseconds to seconds.

It is important to note that other processes besides photon emission
(fluorescence or phosphorescence) also occur when the electron in the
excited state returns back to the ground state. These are, for example, energy
transfer, radiationless dissipation of energy, and intersystem crossing
(Fig. 3).

At this point we would like also to remember that the lifetime (τ) of
the excited state is defined by the average time that the molecule spends in
the excited state prior to return to the ground state and is expressed by
$\tau = 1/\Gamma + K$ where Γ is the emissive rate of the fluorophore and K the rate
of radiationless decay to S_0. The quantum yield of fluorescence is defined
as the ratio of the number of photons emitted to the number absorbed.

These two parameters can be modified by any factors that affect either
of the rate constants, as for example changes in solvent polarity, as we can
see in Table I. Experimentally, the quantum yield can be determined from
measurements of the intensity of the fluorescence and the light absorbed
by the molecules of interest (Weber and Teale, 1957). There are two widely
used methods for the measurement of fluorescence lifetimes. These are the
pulse method and the harmonic or phase-modulation method. In the pulse
method the sample is excited with a brief pulse of light and the time-
dependent decay of fluorescence intensity is measured. In the traditional
harmonic response technique, the sample is excited by light with an intensity
modulated sinusoidally at high frequencies, typically in the megahertz range.
In this case, the fluorescence signal will also be modulated sinusoidally,
but the finite duration of the excited state will lead to a phase delay and

Table I. Spectroscopic Characteristics of 1,8 AEDANS in Solvents with Different Polarities[a]

Solvent system	λ_{abs}^{max} (nm)	$10^{-3}\varepsilon$ (cm^2 mM^{-1})	λ_{em}^{max} (nm)	Quantum yield	τ (n sec) N^2	τ (n sec) Air
Water	342.5	6.3	530	0.045	2.7	2.0
Ethanol–water (20)	345.5	6.8	520	0.073	4.1	4.1
Ethanol–water (40)	346.5	6.8	510	0.120	7.1	6.6
Ethanol–water (60)	347.5	7.3	500	0.19	11.1	10.0
Ethanol–water (80)	348.0	7.5	490	0.29	16.1	12.9
Ethanol	349.0	7.4	480	0.52	25.0	12.6
Propylene glycol	349.0	6.5	488	0.45	19.6	
Dimethylformamide	352.0	7.6	463	0.71	25.8	
Dioxane	352.0		480			

[a] Lifetimes were measured on the phase fluorometer at 28.4 MHz using modulation measurements. Number
in parentheses indicates the percent ethanol in ethanol–water mixtures (from Hudson and Weber, 1973).

demodulation of the fluorescence emission relative to the excitation. Measurement of phase delay and modulation ratio provides independent determinations of the fluorescence lifetime. The very recent appearance of true multifrequency phase fluorometry has enormously extended the scope and power of the harmonic response technique. For more information see the review of Gratton *et al.* (1984).

Because the electron distribution and geometry of a molecule in an electronically excited state are often very different from those on the ground state, the interactions of the excited molecule with the surrounding molecules are often different from the interactions of the ground state molecule. Thus, excited molecules may undergo many kinds of reactions with other ground state molecules.

There are many bimolecular reactions of excited molecules that result in the formation of transient excited complexes called exciplexes.

A simple scheme that permits us to visualize these reactions is

$$A + hV \rightarrow A^*$$
$$A^* + B \rightarrow (AB)^* \xrightarrow{\hspace{2cm}} A + B \hspace{2cm} \text{(Quenching)}$$
$$\textit{exciplex} \xrightarrow{\hspace{1.5cm}} (A^*B \rightleftharpoons AB^*) \rightarrow A + B + \text{fluorescence}$$
$$\text{(Energy transfer)}$$
$$A^- + B^+$$
$$\text{or} \hspace{2cm} \text{(Charge transfer)}$$
$$A^+ + B^-$$

Molecules A and B do not form a complex when both molecules are in the ground state. When one (A) is excited, complex formation with (B) takes place. This excited complex (AB)* is called an exciplex and is stable only in the excited state. The exciplex may relax to the ground state, which is dissociative, giving the original unexcited starting molecules. Many exciplexes (AB)* fluoresce and the spectrum is characteristically shifted to longer wavelengths than the fluorescence from the uncomplexed excited partner (A)*.

In the next sections we will have opportunities to understand and interpret a little more about these processes through the analysis of some results obtained with the Ca^+ATPase.

2.2. Ion Binding and Changes in Intrinsic Fluorescence

Almost all studies relating to the fluorescence of proteins yield information about the conformation of the macromolecules. Since the observed fluorescence of proteins is based on a delicate balance of interacting forces, it is extremely sensitive to changes in the local environment of the aromatic

amino acids residues. Thermal activation, denaturation, ionization, substrates binding, and other types of reaction lead to quenching or increasing
of the fluorescence, changes in fluorescence polarization and changes in
excitation and/or emission spectra.

Careful measurements of the fluorescence as a function of a variable
correlated with these processes reveal transitions that are indicative of
conformational changes that make the aromatic residues more or less
available to solvent or to other functional groups within the molecule.

In this section we will review how the Ca^{2-}ATPase fluorescence
responds to binding of its main ligands: calcium, ATP, Pi, and Mg^{2+}.

Dupont (1976) and Dupont and Leigh (1978) reported that the binding
of Ca^{2+} to the Ca^{2+}ATPase results in an increase of the intrinsic fluorescence,
and this change is accompanied by a slight blue shift of the emission
spectrum. This effect is reversible; the fluorescence intensity decreases to
its original value if, after the addition of Ca^{2+}, an excess of the calcium
chelator EGTA is added to the medium (Fig. 4).

The fact that calcium ions affect the intrinsic fluorescence of the Ca^{2+}
ATPase is a very good indication that binding of this ion is associated with
a conformational change of the protein. The results observed are consistent
with a specific decrease in exposure of the fluorescent tryptophanyl residues
to the polar environment. Using stopped-flow fluorimetry, Dupont and
Leigh (1978) were able to follow the kinetics of the fluorescence change
promoted by the binding and dissociation of Ca^{2+}. The fluorescence intensity
rises slowly after the addition of Ca^{2+} and falls more rapidly when the
added Ca^{2+} is chelated by EGTA (Fig. 5). At 22°C the rate constants of
these two processes are 5 and 30 sec^{-1}, respectively. Both rates were considered to be too slow to represent directly the binding and release of Ca^{2+},
and so a two-step process was proposed. In one of these steps the enzyme
would bind Ca^{2+}, and in the other the enzyme would change its conformation.

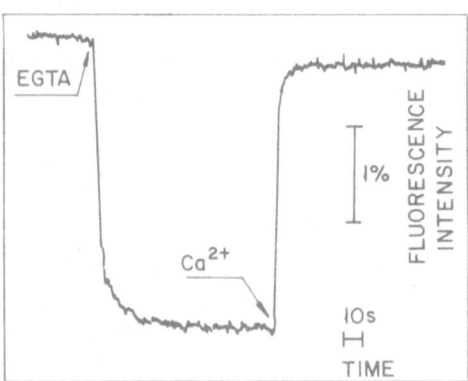

Figure 4. Fluorescence intensity
changes in sarcoplasmic reticulum
ATPase induced by Ca^{2+}. Intrinsic tryptophan fluorescence intensity was
measured at excitation wavelength of
295 nm and emission at 350 nm. Sarcoplasmic reticulum ATPase was suspended in a medium containing 10 mM
TRIS-MOPS   pH 7.2,   100 mM KCl,
5 mM $MgCl_2$,   and   where   indicated
200 μM EGTA or 300 μM Ca^{2+}. (From
Dupont *et al.*, 1985.)

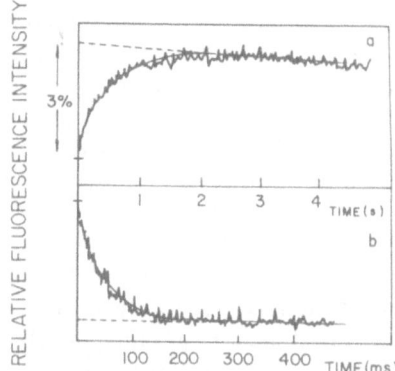

Figure 5. Time resolution of fluorescence changes induced by calcium binding or removal. Sarcoplasmic reticulum ATPase intrinsic tryptophan fluorescence was measured in a stopped-flow apparatus, immediately after the addition of Ca^{2+} (a) or removal of free calcium by the chelator EGTA (b). The temperature of the assay was 11°C. (From Dupont and Leigh, 1978.)

When these fluorescence intensity changes in purified ATPase are studied as a function of the final free Ca^{2+} concentration in the medium (Verjovski-Almeida, 1981), a cooperative pattern appears (Fig. 6). The calcium ion concentration dependence of the fluorescence change indicates an affinity for this ion equivalent to that found by direct titration of the calcium binding sites with ^{45}Ca, i.e., approximately 0.5 μM at pH 7.2. When the lipid membrane vesicles are dissolved with nonionic detergent $C_{12}E_8$ the cooperative pattern of Ca^{2+} binding is lost (Verjovski-Almeida and Silva, 1981) revealing that the detergent has altered the cooperative interaction between the two Ca^{2+} sites in the soluble ATPase enzyme. The detergent solubilized ATPase is unstable in the absence of Ca^{2+} and titration of Ca^{2+} sites by regular equilibration of the soluble enzyme with Ca^{2+} through

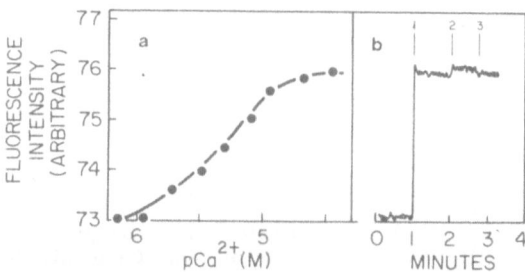

Figure 6. Cooperative calcium binding to sarcoplasmic reticulum ATPase detected by fluorescence changes. (a) Intrinsic tryptophan fluorescence intensity was measured with sarcoplasmic reticulum ATPase suspended in different media containing the free calcium concentrations indicated. (b) Tryptophan fluorescence intensity of ATPase suspended in a medium containing 0.5 mM EGTA was measured as a function of time. At time marked 1, $CaCl_2$ was added to give a total concentration of 0.55 mM $CaCl_2$. At times marked 2 and 3 additional $CaCl_2$ was included to bring the total calcium to 4.55 and 8.55 mM. Micromolar calcium already promotes the maximal fluorescence increase and millimolar calcium causes no further change. (From Verjovski-Almeida, 1981.)

dialysis or chromatography methods are slow processes that lead to partial denaturation of the enzyme. The titration of Ca^{2+} sites in soluble ATPase by fluorescence changes (Verjovski-Almeida and Silva, 1981; Ludi et al., 1982) is an example of application of a unique technique that is both sensitive and fast enough to permit the detection of changes induced by calcium addition or removal from the enzymatic sites before the soluble enzyme becomes inactivated.

Since the first report of an intrinsic fluorescence change induced by calcium binding, all other known substrates have also been reported to affect the intrinsic fluorescence of the Ca^{2+}ATPase. Lacapere et al. (1981) measured the rate of phosphorylation of the enzyme by Pi from the increase in fluorescence that they observed in the absence of calcium at acid pH, a condition in which the Ca^{2+}ATPase can directly be phosphorylate by Pi. ATP or ADP binding in the absence of calcium also produced an important fluorescence increase while in the presence of calcium ATP induced a very fast decrease of the intrinsic fluorescence (Dupont et al., 1985a).

Careful analysis of the intrinsic fluorescence of sarcoplasmic reticulum vesicles under various conditions reveals that different emission spectra are induced by Mg^{2+} or Ca^{2+} ions (Guillain et al., 1984) as we can see in Fig. 7. Mg^{2+} essentially induces a blue shift whereas Ca^{2+} induces a 5% fluorescence enhancement in addition to the blue shift. As a result, the difference emission spectrum for the $^*E \rightarrow E \cdot Ca$ transition is positive for all emission wavelengths whereas the difference spectrum for the $^*E \rightarrow E \cdot Mg$ transition is positive for the shortest wavelengths and negative for higher wavelengths. Therefore, by choosing the appropriate wavelengths it is possible either to study calcium binding without any interference from the Mg^{2+} signal or to

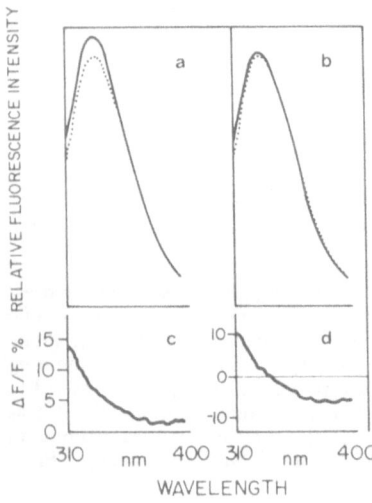

Figure 7. Emission spectra of ATPase in the presence of either Ca or Mg. Intrinsic tryptophan fluorescence emission spectra of sarcoplasmic reticulum ATPase was obtained. Excitation was at 290 nm. The solid lines represent in (a) the addition of 1 mM Ca^{2+} and in (b) the addition of 100 mM $MgCl_2$. The dashed lines were control spectra obtained with ATPase in the presence of 2 mM EGTA. In (c) and (d) are shown the relative fluorescence intensity changes $\Delta F/F$ as calculated from (a) and (b), respectively. (From Guilain et al., 1984.)

cumulate the two effects. Guillain *et al.* (1984) analyzed the interaction of Mg^{2+} with the protein. They showed that magnesium drives the enzyme towards a conformation that binds calcium in a biphasic time course (Fig. 8).

The examples given above illustrate that recording of changes in intrinsic fluorescence provides a good method to study the events associated with conformational changes of the Ca^{2+}ATPase. However, one should have in mind that each different spectral intensity reflects an average conformation that in most cases should not be assigned to only a single specific step in the catalytic cycle of the enzyme. Combination of spectroscopic methods with the actual isolation and characterization of the chemical reactivity of an intermediate species should provide additional information on the different steps of the catalytic cycle of the ATPase.

2.3. Nucleotide Binding Using Fluorescent Analogs

In addition to the study of the intrinsic fluorescence of aromatic amino acids of membrane proteins, one can choose to look at fluorophores that are foreign to the system. These probes can be selected for their improved spectral properties over those of the tryptophan residues and for the possibility of placing the probes at certain specific protein domains, which might cause the fluorescence to be more sensitive to the phenomena one wishes to quantify. Such extrinsic fluorophores can be either covalently attached to amino acids of the protein or covalently linked to the natural ligands which are known to bind reversibly to the protein under study. The former will be discussed in a subsequent section, with the study of ATPase labeled with the sulfhydryl reactive fluorophore pyrene-maleimide. The latter case will be exemplified in this section with the study of binding to the ATPase of the trinitrophenyl-ATP derivative, an analog of the natural substrate ATP.

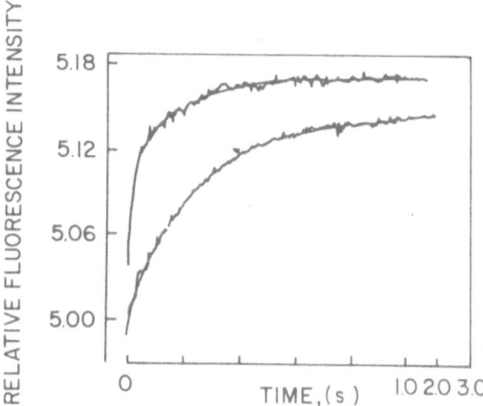

Figure 8. Time-resolved fluorescence changes induced by calcium in the absence and presence of magnesium. Sarcoplasmic reticulum ATPase suspended in media containing EGTA was rapidly mixed with excess Ca^{2+} at time zero and the intrinsic fluorescence changes were measured in a stopped-flow apparatus. The upper trace is for a medium containing no Mg and the lower trace for a medium with 20 mM $MgCl_2$. (From Guillain *et al.*, 1984.)

Other possibilities include the use of ligands such as the metal lanthanides, which are themselves luminescent and have been shown to bind to Ca^{2+} sites of proteins including to the specific high-affinity Ca^{2-} sites of sarcoplasmic reticulum ATPase (Scott, 1985).

In general, the fluorescence emission of a protein-bound ligand is obtained and compared to that of the ligand free in solution; some of the properties of the binding site domains of the protein can be inferred from such studies.

The ATP analog 2′,3′-O-(2,4,6-trinitrophenyl) adenosine 5′-triphosphate (TNP-ATP) has been used for the study of the membrane-bound enzyme (Na-K)ATPase (Moczydlowski and Fortes, 1981) and also in myosin (Hiratsuka and Uchida, 1973) and F_1-ATPase (Grubmeyer and Penefsky, 1981). In sarcoplasmic reticulum ATPase Watanabe and Inesi (1982) found that TNP-ATP binds to the enzyme and that the fluorescent signal of the probe increases upon binding (Fig. 9). A most interesting phenomenon is that, under conditions in which part of the binding sites were occupied by ATP and part by TNP-ATP (Fig. 10), ATP utilization by the enzyme in the presence of Ca^{2+} produces a prominent fluorescence enhancement. Since the total amount of radioactive TNP-ATP bound to the ATPase was *not* found to increase upon addition of ATP, the observed fluorescence enhancement is probably due to a conformational change in the environment of the binding site consequent to phosphorylation of the enzyme by ATP. Therefore TNP-ATP serves as a reporter of an ATPase conformational

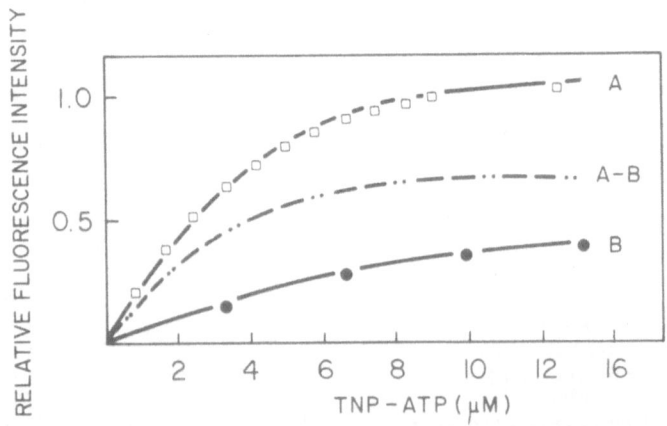

Figure 9. Emission intensities of TNP-ATP fluorescence in the presence (a) and absence (b) of ATPase. The fluorescent ATP analog 2′-3′-O-(2,4,6,-trinitrophenyl)ATP was excited at 410 nm and emission was measured at 515 nm. Different concentrations of TNP-ATP were used, and the intensities of fluorescence emission were recorded in media containing 30 μg/ml sarcoplasmic reticulum ATPase (curve A) or no ATPase (curve B). (From Watanabe and Inesi, 1982.)

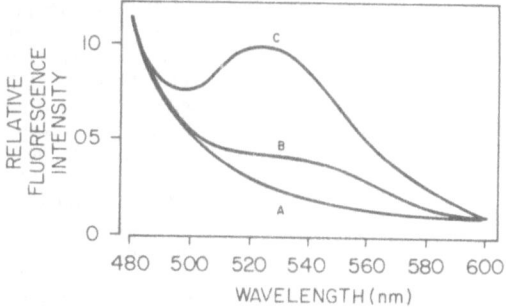

Figure 10.   Fluorescence spectra of bound TNP-ATP in the presence and absence of ATP. Sarcoplasmic reticulum ATPase was mixed with 5 μM TNP-ATP and the emission spectrum of the fluorescent analog of ATP was recorded (B). ATP at a concentration of 50 μM was included and a new spectrum was recorded (C). In (A) a control spectrum of sarcoplasmic reticulum ATPase was obtained. Excitation at 410 nm. (From Watanabe and Inesi, 1982.)

change that is associated to the calcium-dependent phosphorylation of the catalytic site, an important step in the transport cycle that is known to induce calcium translocation (Verjovski-Almeida *et al.*, 1978).

In the absence of Ca^{2+}, the ATPase is not phosphorylated by ATP, and under these conditions enhancement of fluorescence of TNP-ATP bound to the enzyme was not observed. Instead, addition of increasing amounts of ATP in the absence of Ca^{2+} displaced the bound TNP-ATP and promoted a decrease in fluorescence intensity. Dupont *et al.* (1985b) have followed the decrease in TNP-AMPPNP fluorescence upon addition of ATP as an index for measuring the percent of displacement of bound analog induced by ATP (Fig. 11). The titration curves for the competition between TNP-AMPPNP and ATP obtained at pH 6 revealed that at acid pH nearly all bound TNP-nucleotide could be chased by ATP, being compatible with one single family of sites. At pH 7.4 two classes of nucleotide sites could be revealed, as indicated in the figure by the biphasic curve. In addition, it was found that the biphasic curve was only observed when free Mg^{2+} (5 mM) was present. When the Mg chelator EDTA was included or Mg was added in an amount just sufficient to form the ATP-Mg complex and very low free Mg^{2+} was present, the chase of TNP-nucleotide by ATP was a simple sigmoid. From these data and from the titrations of fluorescence chase by ATP at different TNP-nucleotide concentrations, the affinities for ATP at the two classes of sites was estimated to be 3 μM and approximately 1 mM at the high-affinity catalytic site and the low-affinity regulatory site, respectively.

The simplest interpretation of the biphasic chase is that the nucleotide sites of the ATPase may be split into two different classes of sites of very distinct affinities for ATP. Magnesium ions most probably induce a structural

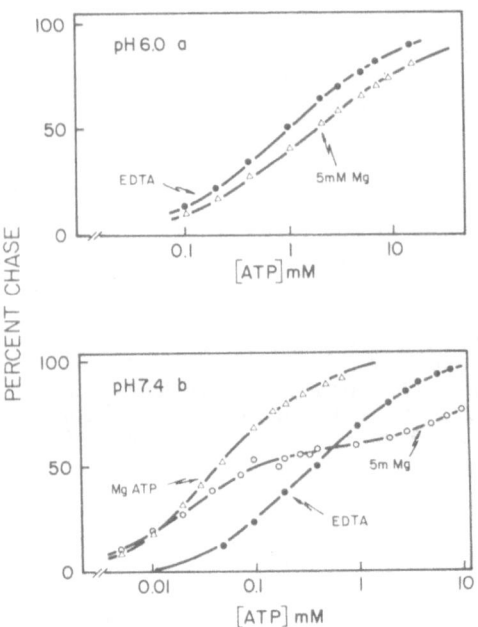

Figure 11. Displacement of bound TNP-ATP by excess ATP. Sarcoplasmic reticulum ATPase was mixed with 4 μM TNP-AMPPNP, a fluorescent analog of ATP, in a medium containing 1 mM EGTA. Fluorescence of bound TNP-nucleotide was measured at 520–570 nm with excitation at 420 nm. Subsequently, ATP was added at the final concentrations indicated and the fluorescence intensities were again recorded. The decrease in fluorescence caused by addition of ATP is expressed as fractional chase; maximal chase corresponds to the maximal decrease in fluorescence and return of the intensity to the level obtained with TNP-AMPPNP and no ATPase (no bound TNP-AMPPNP). Other assay conditions are (a) pH 6.0; (b) pH 7.4; EDTA is data obtained in the presence of 1 mM EDTA; MgCl$_2$ is in the presence of 5 mM free Mg^{2+} (in excess of each ATP concentration); MgATP denotes equimolar concentrations of ATP and MgCl$_2$. (From Dupont *et al.*, 1985.)

change in the ATPase, resulting in a splitting of the ATP binding sites into two distinct families with respect to their binding properties. The use of fluorescent nucleotides for the study of binding sites of low-affinity such as shown in Fig. 11 is a good example of the advantage of employing a fluorescent reporter group that increases its quantum yield upon binding. In such a case, the high concentrations of free TNP-ATP in the medium, which are necessary to achieve saturation of low-affinity sites, will contribute with a much lower background signal than would be the background if other detection methods such as radioactivity were to be used.

The increase in fluorescence yield and the blue shift of emission of bound TNP-nucleotide consequent to a conformational change in the binding site, in this case induced by calcium binding and phosphorylation of the ATPase (Fig. 10), is an example of the general sensitivity of fluorescence emission spectra of fluorophores to the polarity of their surrounding environment. In general, one finds that the emission spectrum shifts to shorter wavelengths (blue shifts) as the solvent polarity is decreased. Conversely, increasing solvent polarity generally promotes shifts to longer wavelengths (red shifts). Red shifts are often accompanied by decrease in the quantum yield of the fluorophore.

The physical origin of the polarity-dependent spectra shifts arises from both the interactions of the dipole moment of the fluorophores with the surrounding solvent and from specific chemical interactions of the fluorophore with the solvent. The emission from a fluorophore occurs at wavelengths that are longer than those of light absorption. The loss of energy between absorption and emission of light, the Stokes' shift, is a result of dynamic processes. These include energy losses due to dissipation of vibrational energy, redistribution of electrons in the surrounding solvent molecules induced by the altered dipole moment of the excited fluorophore (generally increased dipole), reorientation of the solvent molecules around the excited state dipole, and specific interactions between the excited fluorophore and the solvent, such as hydrogen bonding and formation of charge transfer complexes.

Although a complete quantitative separation and description of all these processes is generally not possible when studying emission spectra of fluorophores bound to proteins, the data can often be interpreted in terms of the average environment surrounding the fluorophore.

Thus, the changes in the emission spectra of TNP-nucleotide bound to the ATPase upon phosphorylation of the enzyme are consistent with a change in conformation of the site that leads to a reduced accessibility of polar solvent into the site upon phosphorylation of the enzyme. However, alternative explanations could be given such as a change in the interaction with neighboring quencher amino acids which would lead to formation of a ground state nonfluorescent complex between the probe and the side chains of the amino acid residues. Measurement of lifetime of the probe in the different situations should distinguish between the two possibilities.

2.4. Ion Binding and Change in Solvent Accessibility

Fluorescence quenching has been used to determine the solvent accessibility and fluorescent properties of different fluorophores in proteins (Eftink and Ghiron, 1981). The quencher is in general a small molecule that is in solution and penetrates the protein structure to a certain extent, therefore coming in the vicinity of the fluorophore. The fluorophore can be either an aromatic fluorescent amino acid which is part of the primary structure of the protein or it can be an extrinsic reporter molecule that is covalently linked to the protein one wishes to study. Oxygen is a good quencher of tryptophan fluorescence and it penetrates the protein structure quite freely (Lakowicz and Weber, 1973). Other quenchers that have been used are the negatively charged iodide, the positively charged cesium, and the nonionic acrylamide. In the study of sarcoplasmic reticulum ATPase with the use of fluorescence quenching Kurtenbach and Verjovski-Almeida (1985) were able to show a change in solvent accessibility into certain hydrophobic domains of the ATPase that was induced by binding of Ca^{2+} to the high-affinity Ca^{2+} sites of the enzyme.

Tryptophan fluorescence of the ATPase can be quenched by addition of acrylamide up to 2 M and the data are plotted as F_0/F versus quencher concentration (Fig. 12) where F_0 is the fluorescence intensity in the absence of quencher and F the fluorescence observed at each quencher concentration. The slope of the curve is the Stern–Volmer constant of quenching (k_{SV}).

The Stern–Volmer equation describes the collisional quenching of fluorescence:

$$F_0/F = 1 + kq\tau_0 Q = 1 + k_{SV}Q$$

where k_q is the bimolecular quenching constant, τ_0 is the lifetime of the fluorophore in the absence of quencher, Q the concentration of quencher and $k_{SV} = k_q\tau_0$ is the Stern–Volmer quenching constant. A linear Stern–Volmer plot is generally indicative of a single class of fluorophores, all equally accessible to the quencher. However, a linear Stern–Volmer plot does not prove that collisional quenching of fluorescence has occurred. As we will discuss later, static and dynamic quenching can be better distinguished by lifetime measurements. For the moment it is important to note that dynamic quenching requires contact between the quencher and fluorophore, which in turn requires diffusion of the quencher towards the fluorophore during the lifetime of the excited state. For this reason quenching studies can be used to reveal the localization of fluorophores in proteins and membranes, and to detect changes in the accessibility of fluorophores to quenchers.

In the case of sarcoplasmic reticulum ATPase it was of interest to detect a possible change in the accessibility of tryptophan to acrylamide

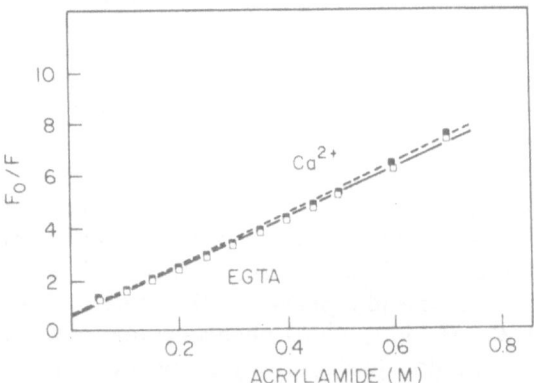

Figure 12. Acrylamide quenching of intrinsic tryptophan fluorescence of ATPase. Sarcoplasmic reticulum ATPase was suspended in media containing either 0.5 mM EGTA or 0.5 mM EGTA plus 0.6 mM $CaCl_2$. The fluorescence intensity of tryptophan emission was recorded (F_0) and increasing concentrations of acrylamide quencher were added as indicated. The intensities in the presence of quencher (F) were recorded. The data are shown in a Stern–Volmer plot, as indicated in the text.

quenching upon binding of Ca^{2+}, the ion that is transported by the enzyme. A change in conformation of the enzyme had already been shown to occur upon binding of calcium (see Inesi, 1985) and a possible change of solvent accessibility associated to this conformational change could be examined by fluorescence quenching. Figure 12 shows however that tryptophan fluorescence quenching was insensitive to Ca^{2+} binding to the high affinity sites of the ATPase, as already reported by Murphy (1978) and Lüdi and Hasselbach (1983).

Because of the large number of tryptophanyl residues of the ATPase (18-19 mol per mol of enzyme) it would in fact be quite unlikely that quenching of total tryptophan emission should be sensitive to the local conformational changes of the protein in the vicinity of the sites that are affected by binding of Ca^{2+}. This is a situation in which the search for an extrinsic probe placed in a more favorable specific domain would be advantageous.

Kurtenbach and Verjovski-Almeida (1985) have succeeded in covalently attaching to the ATPase a thiol-reactive pyrene-maleimide fluorescent probe at a favorable site in a limiting stoichiometric amount. Figure 13 shows that quenching of pyrene-ATPase fluorescence by acrylamide or by cesium is sensitive to the binding of Ca^{2+} to the ATPase. It can be seen that the changes induced by Ca^{2+} binding to the high-affinity sites involve an increase in the Stern–Volmer quenching constant, indicating an increased exposure of hydrophobic domains of the ATPase to the medium solvent.

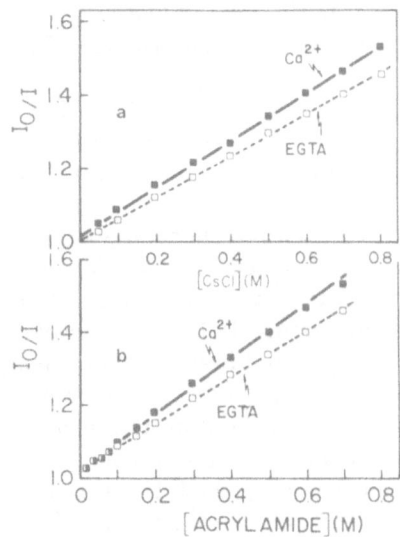

Figure 13. Quenching of pyrene-ATPase fluorescence. Sarcoplasmic reticulum ATPase was covalently labeled with pyrene-maleimide and pyrene fluorescence of the labeled enzyme (F_0) was recorded at 392 nm with excitation at 343 nm. Addition of CsCl (a) or acrylamide (b) resulted in lower fluorescence intensities (F). The data are shown as F_0/F in a Stern–Volmer plot. The media contained either 0.5 mM EGTA (open circles) or 0.5 mM EGTA plus 0.6 mM $CaCl_2$ (solid circles). (From Kurtenbach and Verjovski-Almeida, 1985.)

For pyrene-maleimide-cystein in solution the Stern–Volmer fluorescence quenching constant with acrylamide is $5.5\ M^{-1}$, and the measured fluorescence lifetime of the adduct is 84 nsec (Kurtenbach and Verjovski-Almeida, 1985) giving a second-order rate constant for acrylamide quenching of $6.6 \times 10^7\ M^{-1}\ sec^{-1}$. The second-order rate constant for quenching of pyrene fluorescence by acrylamide when the fluorophore is on the ATPase taken from the data of Fig. 13 is $4.4 \times 10^6\ M^{-1}\ sec^{-1}$ in the absence of calcium and $5.3 \times 10^6\ M^{-1}\ sec^{-1}$ when Ca is bound to the ATPase sites. The constant is one order of magnitude lower when the fluorophore is on the protein as compared to pyrene-maleimide-cystein in solution, which indicates that the fluorophore is considerably shielded from the medium solvent and is probably attached to a hydrophobic domain on the ATPase.

A change in solvent accessibility into specific sites on proteins can only be accessed by quenching if dynamic quenching is the predominant mechanism, whereby the quencher diffuses into the vicinity of the excited fluorophore. Quenching through this mechanism shortens the fluorescence lifetime of the probe, and measurements of lifetimes in the absence and presence of quencher should distinguish between dynamic and static quenching. In the latter, a complex is formed between the quencher and fluorophore in the ground state, resulting in a nonfluorescent complex. When the complex absorbs light it immediately returns to the ground state without emission of a photon. The observed fluorescence intensity decrease as a function of increasing quencher concentrations in this case reveals the association constant for the nonfluorescent complex formation. The fraction of fluorophore that remains uncomplexed will fluoresce with an unmodified lifetime.

In the Ca^{2+}-ATPase the quenching of pyrene-ATPase fluorescence by acrylamide was shown to be of a dynamic nature (Kurtenbach and Verjovski-Almeida, 1985). Table II shows that the lifetime decreased with increasing acrylamide concentrations. The measurement of fluorescence lifetime is the most definitive method to distinguish between static and dynamic quenching, but when this method is not available one can study the dependence of the quenching constant as a function of temperature.

Dynamic quenching is a diffusive process and diffusion increases with increasing temperature. As a result, the bimolecular quenching constant is expected to increase with increasing temperature. In contrast, increased temperature is likely to result in decreased stability of complexes, and therefore to lower the static quenching constants. When using the quenching methods to infer the solvent accessibility or the exposure of a certain fluorophore in a protein it is important to rule out the possible unwanted formation of ground-state complexes between the quencher and the fluorophore, which can arise from the presence of certain "sites" or pockets on the protein in the vicinity of the fluorophore that trap the quencher and

Table II. Fluorescence Lifetimes of
Pyrene-ATPase Emission[a]

[Acrylamide]	τ (nsec)
0	144
0.3 M	142
0.6 M	115
0.9 M	89
$0 + C_{12}E_8$	149

[a] Sarcoplasmic reticulum ATPase was covalently labeled with pyrene-maleimide and the lifetime of pyrene-ATPase emission was measured on the phase fluorometer using modulation measurements at 6 MHz. The sample contained increasing concentrations of acrylamide as indicated; in one sample acrylamide was omitted and the nonionic detergent $C_{12}E_8$ was included. (From Kurtenbach and Verjovski-Almeida, 1985.)

prevent it from homogeneously diffusing into and out of the protein domain to be studied.

Under certain conditions both dynamic and static quenchings of fluorescence occur, giving rise to a nonlinear upward curved Stern–Volmer plot, concave towards the y axis. The dynamic portion of the observed quenching can be determined separately by lifetime measurements, and the contribution of each component can be evaluated. Another situation encountered is the quenching of a fraction of the total fluorescence, which is generally characterized by downward curving Stern–Volmer plots. This arises from the existence of two populations of fluorophore in the sample, one of which is inaccessible to quencher.

The interior region of proteins and the membrane-protein interface are nonpolar and one may expect that at least some of the tryptophan residues of a protein are in these regions and therefore are inaccessible to quenchers in the external aqueous phase. In fact, in sarcoplasmic reticulum ATPase it was found that about 54% of the tryptophan residues were accessible to the water soluble quencher N-methyl-picolinium (Fig. 14), giving rise to downward curved Stern–Volmer plots (Shinitzky and Rivnay, 1977) which are linearized by a modification of the Stern–Volmer plot (Lehrer, 1971).

A different approach is to use fluorescent quenchers that are dissolved in the lipid membrane and will eventually come into the vicinity of the lipoprotein microenvironment through the lipid–protein interface. In the Ca^{2+}-ATPase quenching of tryptophan intrinsic fluorescence was obtained by using spin-labeled phospholipids that were incorporated into native

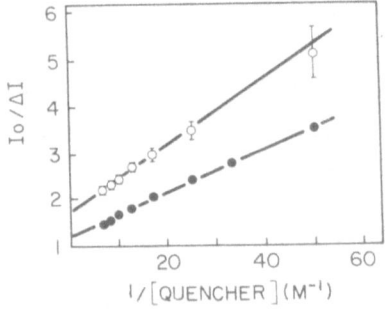

Figure 14. Modified Stern-Volmer plot for quenching of tryptophan fluorescence. Sarcoplasmic reticulum ATPase was suspended in a buffer medium and tryptophan fluorescence emission was recorded (F_0). Quencher methyl picolinium was added and the decrease in fluorescence intensities were obtained (ΔF). The data are plotted as $F_0/\Delta F$ versus the inverse of quencher concentration; a modified Stern-Volmer plot as described in the text. The solid circles represent ATPase inserted in the lipid membrane. The intercept near 2 indicates partial exposure of the tryptophanes to the quencher. The open circles represent ATPase in a medium containing trifluorethanol which disrupted the membrane and exposed a large fraction of fluorophores the quencher. (From Shinitzky and Rivnay, 1977.)

vesicles (London and Feigenson, 1981). The spin-label nitroxide moiety of the phospholipids is a quencher of tryptophan fluorescence, and it was possible to show that 70%–80% of the fluorescence could be quenched by these phospholipids (London and Feigenson, 1981). Using lipid-soluble fluorescent antibiotic X-537A (Verjovski-Almeida, 1981) it was possible to show that 70% of the intrinsic tryptophan fluorescence was quenched by this lipid soluble probe. The quenching mechanism in this case involved radiationless energy transfer from excited tryptophan to the probe fluorophore. In both cases, with spin-labeled phospholipids and with lipid-soluble X-537A the partial quenching indicates that the majority of the tryptophanyl residues (70%–80%) are accessible from the lipid–protein interface microenvironment, but a second population of fluorescent residues is present that is either in the interior of the protein or in a region more accessible from the aqueous medium.

3. MACROMOLECULAR ASSOCIATION

3.1. Polarization of Fluorescence of Long-Lived Fluorophores

Fluorescence emission is polarized when the excitation light is polarized. This polarization is a result of the photoselection of fluorophores according to their orientation relative to the direction of the polarized excitation. A fluorophore can be considered to have an oscillating dipole moment. We will refer to the transition moments for excitation and emission as the absorption and emission dipoles. For a fluorophore in solution the absorption and emission dipoles will be randomly oriented in an isotropic and homogeneous solvent. Excitation with polarized light selects for a range of orientations, resulting in a population of excited fluorophores that are symmetrically distributed within a small angle around the axis of the polarized excitation. The resulting emission of fluorescence is polarized, and for colinear absorption and emission dipoles the possible maximum polarization of fluorescence of 0.5 is obtained, when there are no further processes such as rotational diffusion, energy transfer, or scatter that result in depolarization.

For a fluorophore in dilute solution that undergoes negligible Brownian motion between excitation and emission, the polarization of fluorescence is a maximum called the fundamental or limiting polarization P_0. Rotational diffusion of the fluorophore in solution during the lifetime of the excited state results in a certain degree of depolarization of the emission. Therefore, measurement of the polarization of emission of a sample can reveal the rotational rate of the fluorophore, which in turn is dependent upon the viscous drag imposed on the fluorophore by the solvent. One can estimate the rotational rate by changing solvent viscosity and observing the resulting change in fluorescence polarization. The first biochemical application of this technique is found in the pioneering work of Weber (1952) on the estimation of the rotational rates of proteins.

The estimation of the internal viscosities of membranes and the effects of membrane composition on their phase transitions have also been extensively studied by polarization with the use of lipophilic fluorescent probes such as diphenyl hexatriene (DPH) (Shinitzky *et al.*, 1971; Lentz, 1979).

The measurement of fluorescence polarization is illustrated in **Fig. 15**. The sample is excited with vertically polarized light. The electrical vector

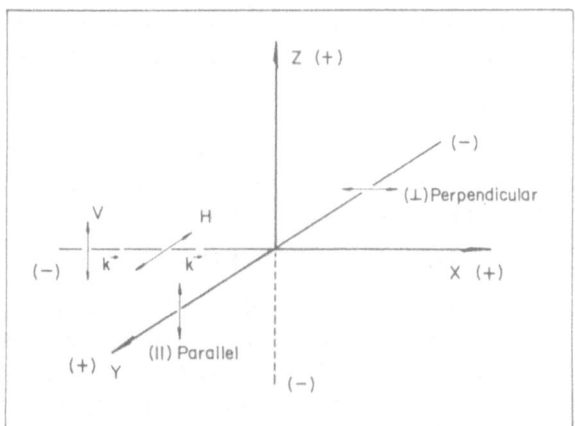

Figure 15. Diagram of reference axis in fluorescence polarization measurements. The excitation light travels along the axis and V and H represent directions of vertically and horizontally polarized light. The emission is detected along the y axis and ($\perp$) and ($\parallel$) refer to directions perpendicular and parallel to V, respectively.

of the excitation is parallel to the z axis. The intensity of the emission is measured through a second polarizer. When the emission polarizer is parallel ($\|$) to the direction of the polarized excitation the observed intensity is $I_{\|}$. When the polarizer is perpendicular ($\perp$) to the excitation the intensity is $I_{\perp}$. The polarization (P) is

$$P = \frac{I_{\|} - I_{\perp}}{I_{\|} + I_{\perp}}$$

For completely polarized light $I_{\perp} = 0$ and $P = 1.0$. This high value is obtained when excitation scattered light is detected. Scattered light is an important source of artifacts when measuring the polarization of fluorescence emission of low intensities; emission spectra of the samples should be obtained in order to make sure that at the wavelengths used for the detection no light scattered from the excitation beam is being collected. Both Rayleigh and Raman scatter must be considered. The latter has a characteristic pattern for each solvent used and in water the Raman line is at 3600 cm^{-1} lower wave number than the exciting light. It should not be mistaken for fluorescence emission. Values of polarization of fluorescence of fluorophores in solution are never equal to 1 owing to the angular dependence of photoselection, which lowers the polarization of the resulting fluorescence emission. Measured polarization for a randomly oriented sample greater than 0.5 indicates the presence of scattered light in addition to fluorescence.

Rotational diffusion of the fluorophore is the predominant cause of fluorescence depolarization, and if Brownian motion of the molecules is very high the initial orientation is lost in the few nanoseconds between excitation and emission. In dilute solutions the polarization is a good indicator of the motion of the particles, and the polarization observed is determined by three factors: the limiting polarization P_0 that would be obtained if no disorientation followed the excitation, the rate of rotation R of the particles (rotational diffusion coefficient), and the time interval between excitation and emission τ (i.e., the fluorescence lifetime of the fluorophore). Perrin (1926) showed that these quantities are related by

$$1/P - \tfrac{1}{3} = (1/P_0 - \tfrac{1}{3})(1 - \tau 6R)$$

If the fluorescent molecule is attached to a macromolecule, the shape of it might be quite irregular, and the motion might not be described by a single rate of rotation. However, if as a first approximation the particle is assumed to be spherical, the rate of rotation of a sphere (the rotational diffusion coefficient) is equal to

$$R = \tfrac{1}{6}(R_g T / \eta V)$$

where R_g is the molar gas constant, T is the absolute temperature, η, is the viscosity of the solvent in which the molecule moves, and V is the volume of the rotating particle. Combination of the above two equations gives

$$1/P - \tfrac{1}{3} = (1/P_0 - \tfrac{1}{3})(1 + \tau R_g T/\eta V) = (1/P_0 - \tfrac{1}{3})(1 + \tau/\phi)$$

The rotational time ϕ is $\eta V/R_g T$. The Perrin equation describes the depolarization due to the Brownian motion of a spherical molecule. Temperature and viscosity of the medium can be varied. A plot of $(1/P - \tfrac{1}{3})$ against T/η should give a straight line with intercept $(1/P_0 - \tfrac{1}{3})$ and slope $(1/P_0) - (\tfrac{1}{3})R\tau/V$.

If the molecule volume is known, τ may be determined. If τ is known, the molecular volume or an effective volume in solution may be calculated. When the macromolecule associates into higher oligomeric structures, the state of association under a given condition may be characterized. A variety of solvent and solute media may be studied, as well as a range of temperature and macromolecule concentrations.

It should be noted that the fluorescence lifetime τ must be in the range of the expected rotational correlation time ϕ of the macromolecule to be studied. A lifetime too short compared to ϕ would result in emission much before the rotation has caused a significant reorientation of the excited population leading to $P = P_0$. On the contrary, if τ is too long compared to ϕ the emission will be totally depolarized and $P = 0$. Knopp and Weber (1969) have introduced the use of a pyrenebutyryl probe as a good covalent label of proteins, which has a lifetime of 100-150 nsec and permits the estimation of the molecular volume of proteins up to molecular weights of 10^6 (rotational correlation times of 300-400 nsec).

In sarcoplasmic reticulum ATPase the fluorescence polarization of a pyrene probe covalently attached to the enzyme was measured by Kurtenbach and Verjovski-Almeida (1985). In Fig. 16 a Perrin plot of the data obtained with sarcoplasmic reticulum ATPase inserted in its native lipid membrane vesicles show that polarization was not sensitive to the slow rotational diffusion of the large lipid vesicles (approximately 1000-2000 Å in diameter). As expected, the rotational rate of such a large vesicle is too slow and is outside the range that would be detected with polarization of

Figure 16. Perrin plot of pyrene-ATPase fluorescence polarization. The polarization of pyrene fluorescence emission (p) was measured in a sample of ATPase covalently labeled with a pyrene-maleimide probe. The viscosity of the medium (η) was increased by addition of sucrose up to 70% (w/v). The data were obtained for pyrene-labeled ATPase inserted in its native membrane vesicles (absence of detergent, triangles) and for the same enzyme solubilized by nonionic detergent $C_{12}E_8$ (squares). (From Kurtenbach and Verjovski-Almeida, 1985.)

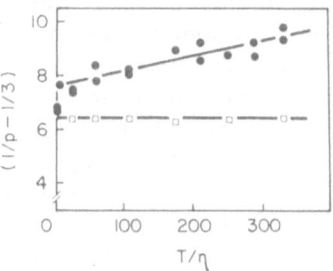

pyrene emission. In previous work with pyrene-labeled ATPase (Lüdi and Hasselbach, 1983) attempts were made to measure polarization of fluorescence emission with ATPase inserted in the membrane vesicles. However, the measurements were done in the presence of a high level of Raman excitation light scatter, which produced a characteristic and intense band in the emission spectrum that was mistaken for an unknown fluorophore of the ATPase.

Polarization of fluorescence is a useful tool to study the degree of association of membrane proteins after solubilization. Addition of nonionic detergent dodecyl octaethylene glycol monoether ($C_{12}E_8$) to sarcoplasmic reticulum vesicles dissolve the ATPase and preserves its activity. The solubilized ATPase retains a high tendency to self-associate into the dimeric state as determined by gel filtration in the presence of high concentrations of detergent (Silva and Verjovski-Almeida, 1985) and measurements of the effective volume of the solubilized ATPase by fluorescence polarization of solubilized pyrene-labeled ATPase (Fig. 16) provides additional support for the dimeric state of association of the protein. Polarization of pyrene-ATPase emission was 0.15 in the vesicular ATPase and it decreased to 0.10 upon addition of detergent. The depolarization reflects the increase in rotational mobility of the ATPase by solubilization and dispersion of the enzyme in the solvent. The polarization of solubilized pyrene-ATPase emission increased with the increase in viscosity of the medium by addition of sucrose up to 70% (Fig. 16) and the Perrin plot was linear. Increase in viscosity by addition of sucrose was preferred in place of a change in temperature, owing to the strong dependence of detergent micelle size upon temperature. It should be preferable in principle to change viscosity by addition of sucrose rather than to change temperature whenever membrane proteins are dissolved in detergent.

The rotational correlation time was obtained from the slope and intercept of Fig. 16 and by a separate determination of τ, the lifetime of pyrene fluorescence in the presence of detergent. The lifetime was measured in the frequency domain by phase-modulation detection (Spencer and Weber, 1970; Gratton et al., 1984) and it was found to be 149 nsec (Table II). The rotational correlation time of the solubilized pyrene-labeled ATPase obtained from Fig. 16 was 532 nsec. Assuming the detergent-protein particle to be a sphere, the volume and the radius can be calculated from the rotational correlation time; the obtained value is 87 Å radius. This value is compatible with the Stokes radius of dimeric solubilized ATPase, which was shown to be 60 Å (Silva and Verjovski-Almeida, 1983) when measured by column chromotography. In general, the obtained values of rotational correlation time are higher than those expected for a typically hydrated spherical model protein (Yguerabide et al., 1970). The larger observed values are probably a result of the nonspherical shape of most proteins.

3.2. Time-Resolved Decay of Polarization

Rotational correlation time of a fluorescent molecule in solution can be obtained by measuring the polarization of the steady-state fluorescence emission and its dependence on the viscosity of the medium, as described in the preceding section. An alternative is to measure the time-resolved anisotropy decay of fluorescence emission after the fluorophore is excited with a flash of polarized light. It should be noted that anisotropy is defined by

$$r = \frac{I_\| - I_\perp}{I_\| + 2I_\perp}$$

and that polarizations (P) and anisotropies (r) can be interchanged using the following relations:

$$P = \frac{3r}{2 + r}, \qquad r = \frac{2P}{3 - P}$$

For a molecule with a single rotation rate (rotation is isotropic and unhindered) the fluorescence anisotropy (r) decays exponentially according to

$$r(t) = r_0 e^{-t/\phi} = r_0 e^{-6Rt}$$

where ϕ is the rotational correlation time of the fluorophore, R is its rotational rate $(6R = \phi^{-1})$ and r_0 is the anisotropy observed in the absence of rotational diffusion. The time-resolved decays of individual polarized components of the emission are

$$I_\|(t) = (\tfrac{1}{3})I_0(t)[1 + 2r(t)]$$

$$I_\perp(t) = (\tfrac{1}{3})I_0(t)[1 - r(t)]$$

At $t = 0$ following a short pulse of exciting light the intensities of the parallel and perpendicular components are given by

$$I_\|(t = 0) = (\tfrac{1}{3})I_0(1 + 2r_0)$$

$$I_\perp(t = 0) = (\tfrac{1}{3})I_0(1 + r_0)$$

At times longer than $t = 0$ the initial difference in intensity between $I_\|(t)$ and $I_\perp(t)$ decays owing to both rotational diffusion of the fluorophore and to the decay of the total emission. For a simple exponential decay of anisotropy, a plot of $\ln r(t)$ versus t is linear, with a slope equal to ϕ^{-1}.

In solubilized sarcoplasmic reticulum ATPase Yamamoto et al. (1984) have measured the rotational correlation time of the ATPase-detergent particle by time-resolved anisotropy decay. The authors used pyrene-maleimide labeled ATPase. Figure 17 shows the results obtained with solubilized labeled ATPase eluted at different fractions from a gel filtration

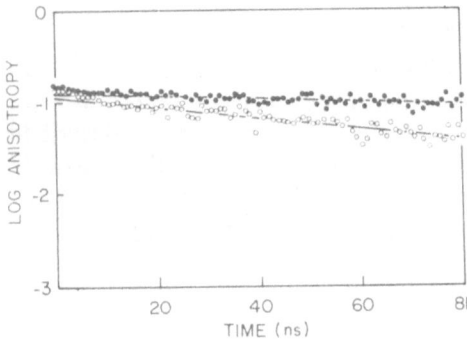

Figure 17. Time-resolved anisotropy decay of pyrene-ATPase fluorescence. Sarcoplasmic reticulum ATPase was covalently labeled with a pyrene-maleimide probe, followed by solubilization in nonionic detergent $C_{12}E_8$ and separation into monomeric (circles) and dimeric (triangles) soluble ATPase by gel filtration. Anisotropy decay of pyrene fluorescence emission was measured after a short pulse of excitation at 351 nm. The lines are the fittings for data points taken at times longer than 15 nsec. The rotational correlation times are 251 nsec for dimeric soluble ATPase (triangles) and 74 nsec (circles) for monomeric. (From Yamamoto *et al.*, 1984.)

column. Two peaks of protein corresponding to dimeric and monomeric solubilized ATPase were obtained (Yamamoto *et al.*, 1984) in agreement with previously reported data (Silva and Verjovski-Almeida, 1983). The anisotropy decay of pyrene fluorescence gave linear plots (Fig. 17) in both dimeric and monomeric solubilized pyrene-labeled ATPase; the rotational correlation times obtained from the slopes were 251 and 74 nsec for dimeric and monomeric ATPase, respectively. These correlation times correspond to the rotation of spherical particles with radius of 80 and 54 Å for the dimer and monomer, respectively. The radius of 54 Å for the monomeric solubilized ATPase is in good agreement with the radius of 47 Å obtained by equilibrium sedimentation or gel filtration studies (Dean and Tanford, 1978; Silva and Verjovski-Almeida, 1983).

The utilization of time-resolved polarization decay has the advantage over steady-state measurements of allowing identification of complex decay patterns, revealing possible anisotropic or hindered rotations. In addition, fast rotating contaminants can be excluded when analyzing the rotation of macromolecules. In Fig. 17, it can be seen that the initial rapid decaying component (within the first 15 nsec) was present, probably representing a population of fast rotating contaminant. For the fittings, data obtained only at times longer than 15 nsec were used.

3.3. Pressure-Induced Dissociation of Macromolecules

Polarization of fluorescence emission is a unique sensitive spectroscopic method to directly detect changes in rotational rate of macromolecules upon dissociation of macromolecular aggregates. On the other hand, hydrostatic pressure is a reversible perturbing variable that is most useful for understanding physical and chemical processes in biological systems where macromolecular interactions are present. Effects of pressure-induced dissociation of macromolecular assemblies have been indirectly examined by the loss in enzymatic activity that accompanies the dissociation of multisubunit enzymes (Heremans, 1982). Direct observation of dissociation has been obtained by combination of hydrostatic pressure and fluorescence spectroscopy in the work of Paladini and Weber (1981), who measured the changes in polarization of fluorescence emission of enolase under pressure, showing that pressure promoted dissociation of dimeric enolase into monomers.

Verjovski-Almeida et al. (1986) have recently extended the observations originally obtained for water-soluble enolase into the study of the sarcoplasmic reticulum ATPase solubilized in detergent. Figure 18 shows the polarization of fluorescence of pyrene-labeled solubilized ATPase as a function of hydrostatic pressure up to 2 kbar. It is shown that pressure induces a reversible decrease in polarization. The decrease in polarization from 0.174 at atmospheric pressure to 0.128 at 2–2.2 kbar is compatible with dissociation of dimeric ATPase into monomers.

Separate measurements at atmospheric pressure of the polarization of fluorescence of pyrene butyryl-ATPase as a function of viscosity (Verjovski-Almeida et al., 1986) gave a rotational correlation time of 380 nsec for the solubilized ATPase, corresponding to rotation of a spherical particle of 75 Å radius (dimeric ATPase). At 2 kbar the polarization obtained (0.128) would correspond to a decrease in the correlation time to approximately 110 nsec, which would represent the rotation of a particle of 50 Å radius.

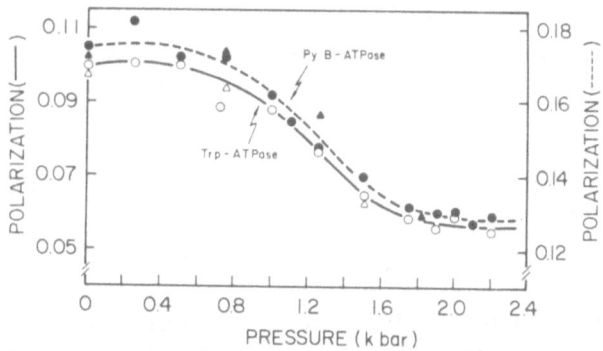

Figure 18. Effect of pressure on the polarization of ATPase intrinsic fluorescence and pyrene-butyryl-ATPase fluorescence. Sarcoplasmic reticulum ATPase (closed symbols) or ATPase covalently labeled with a pyrene probe (open symbols) was solubilized in nonionic detergent at a protein concentration which yields predominantly dimeric ATPase. Polarization of fluorescence of intrinsic tryptophans (Trp-ATPase, closed symbols) or of pyrene-ATPase (Pyb-ATPase, open symbols), was recorded under hydrostatic pressure, in the range indicated. The circles are the polarization obtained upon increase in pressure from atmospheric pressure. The triangles are the polarization recovered after reduction of the pressure from 2.2 kbar. (From Verjovski-Almeida et al., 1986.)

This is compatible with the predominance at 2 kbar of monomeric solubilized ATPase.

Figure 18 shows that pressure dissociation could be detected not only by a change in polarization of pyrene emission but also by a change in polarization of the intrinsic tryptophan fluorescence emission of the solubilized ATPase. The absolute values of polarization were different, but the pressure profiles were identical for both pyrene and tryptophan emission. The pressure required for half-dissociation was in both cases 1.1 kbar. Tryptophan fluorescence lifetime is considerably short (2–4 nsec) and the polarization changes are probably reflecting changes in local mobility of tryptophan residues as the dimeric ATPase dissociates, rather than reflecting changes in the overall rotational motion of the macromolecule, which was only detected by the long-lived fluorescence of pyrene butyryl-ATPase (120 nsec). For enolase Paladini and Weber (1981) also found that tryptophan residues were subjected to changes of local mobility as a function of dissociation, and that tryptophan fluorescence polarization could be used to report the pressure-induced dissociation of enolase.

4. PROTEIN–PROTEIN AND PROTEIN–LIPID INTERACTIONS

Fluorescence energy transfer is the transfer of the excited state energy from a donor to an acceptor. This transfer is a nonradiative process and is primarily a result of dipole–dipole interaction between the donor and the acceptor. The efficiency of transfer is dependent upon the separation distance between the donor and acceptor chromophores, the extent of overlap between the emission spectrum of the donor and the absorption spectrum of the acceptor, and the relative orientation of the donor and acceptor transition dipoles.

The rate of energy transfer from a specific donor to a specific acceptor (k_t) is given by

$$k_t = (1/\tau_d)(R_0/r)^6$$

where τ_d is the lifetime of the donor in the absence of acceptor, r is the distance between the donor and acceptor, and R_0 is a characteristic distance called the Förster distance at which the efficiency of transfer is 50%. Efficient transfer of energy in most donor–acceptor pairs occurs at separation distances between 20 and 50 Å.

Fluorescence energy transfer measurements have been used to determine distances between donors and acceptors. These chromophores can be binding, for example, at active sites on different catalytic subunits, and in this way one can determine the distance between catalytic sites of different subunits, as can be seen in the work of Hahn and Hammes (1978) in the study of aspartate transcarbamylase.

Vanderkooi *et al.* (1977) and Yantorno *et al.* (1983) used the energy transfer measurements to obtain information on the association of Ca^{2+}ATPase from sarcoplasmic reticulum within the plane of the membrane vesicles. Figures 19 and 20 show the results obtained by Yantorno *et al.* (1983). Solubilized ATPase was labeled separately with either *N*-(1-pyrene)-maleimide (NPM) the donor, or *N*-(7-dimethylamino-4-methyl-3-coumarinyl) maleimide (DACM), the acceptor. The good overlap between the spectra of fluorescence emission of NPM and of excitation of DACM ensured that they are a good donor–acceptor pair for energy transfer (Fig. 19).

When the separately labeled ATPase were mixed together in the presence of $C_{12}E_8$ and the detergent was removed by incubation with Bio-Beads SM-2, the two types of labeled ATPase were inserted together into reconstituted membrane vesicles and a significant amount of fluorescence energy transfer was observed between NPM-ATPase and DACM-ATPase (Fig. 20). However, energy transfer did not occur when each type of labeled ATPase was reconstituted separately into distinct vesicles (in separate tubes) and the vesicles were simply mixed in the same solution as a control (Fig. 20).

Yantorno *et al.* (1983) also measured the fluorescence lifetime of the donor NPM-ATPase in the presence and absence of the energy acceptor DACM-ATPase and showed that the fluorescence lifetime of the donor was significantly shortened by the presence of the energy acceptor, which is

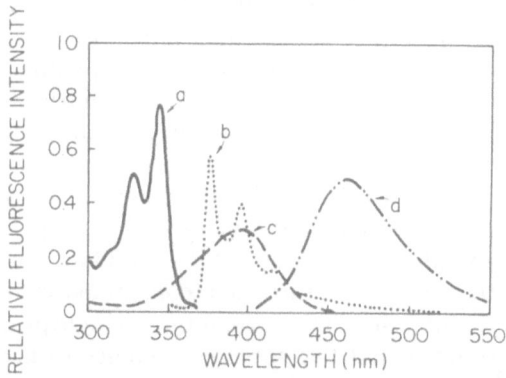

Figure 19. Emission and excitation spectra of ATPase covalently labeled separately with either a pyrene probe or a dimethylaminomethyl coumarinyl probe. Sarcoplasmic reticulum ATPase was labeled with either pyrene-maleimide (spectra a and b) or with dimethyl amino methyl coumarinylmaleimide (spectra c and d). The excitation spectra are in solid lines (a and c) and the emission spectra in broken lines (b and d). The overlap of spectra b and c characterizes the two probes as a good pair of donor acceptor for energy transfer. (From Yantorno *et al.*, 1983.)

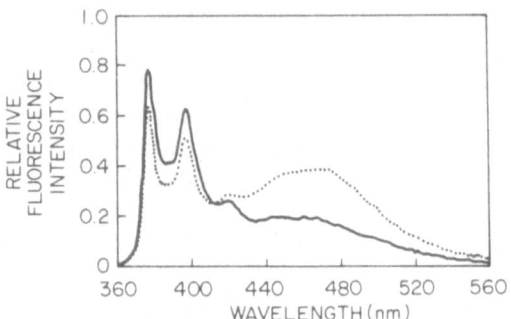

Figure 20. Energy transfer between pyrene-ATPase and coumarinyl-ATPase. Sarcoplasmic reticulum ATPase was separately labeled either with pyrene maleimide (pyrene-ATPase) or with dimethyl amino methyl coumarinyl maleimide (DACM-ATPase). Fluorescence emission spectra were obtained upon excitation of pyrene at 343 nm. The solid line is a control with each labeled ATPase incorporated into separate membrane vesicles and simply mixed in the test tube. The dashed line is the spectrum after the two different labeled ATPase were incorporated together into the same lipid membrane. (From Yantorno et al., 1983.)

evidence of the fluorescence energy transfer between NPM-ATPase and DACM-ATPase. They concluded that an oligomeric form of the ATPase molecules and not a random arrangement was formed within the plane of reconstituted membranes after $C_{12}E_8$ was removed from the mixture. These experiments illustrate how energy transfer measurements can be used to examine protein–protein interactions.

The energy transfer can also be used to assess the association of the ATPase protein with the lipid membrane. Vaz et al. (1977) used as donors the ATPase tyrosine or tryptophan residues and a lipid acceptor having an absorption range in the 300–380 nm, the DNS-kephalin. They have also used the method to quantify the association between trypsin and MPA and also detecting association of melittin and acetylcholine esterase with MPA and lecithin vesicles.

A different approach was used by Gomez-Fernández et al. (1985), who decided to study the protein–lipid interactions using the fluorescence quenching of tryptophanyl residues of Ca^{2+}ATPase by acrylamide at different temperatures (Fig. 21). Enzyme reconstituted with synthetic phospholipid was used to obtain systems containing a single protein and a single phospholipid in order to test the possible influence of membrane fluidity on the accessibility of tryptophan residues.

As we can see in Fig. 21 the acrylamide quenching constants (k_{SV}) show a sharp increase in samples of ATPase reconstituted with dimyristoylphosphatidylcholine (DMPC) and dipalmitoylphosphatidylcholine (DPPC) at temperatures slightly below the T_c transition temperature of the pure phospholipid. The transition in k_{SV} observed with the reconstituted

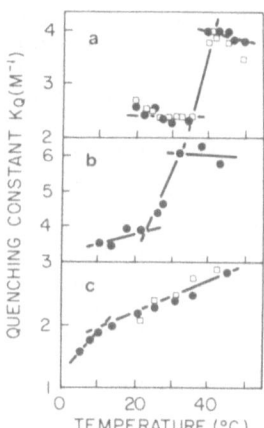

Figure 21. The influence of different lipid compositions on the quenching of intrinsic ATPase fluorescence by acrylamide. Sarcoplasmic reticulum ATPase was reconstituted into lipid vesicles prepared with (a) dipalmitoyl phosphatidyl choline, (b) dimiristoyl phosphatidyl choline, and (c) either endogenous sarcoplasmic reticulum lipids or egg yolk lecithin. The quenching of fluorescence of intrinsic tryptophanyl residues of each ATPase preparation was measured in the presence of variable concentrations of acrylamide up to $2M$ and the Stern–Volmer quenching constants (K_0) were obtained at the different temperatures indicated. (From Gomez-Fernandez *et al.*, 1985.)

samples, at 37 and 21°C for DPPC and DMPC-ATPase compared with 41 and 23°C for pure DPPC and DMPC are attributed to changes in protein tryptophanyl's accessibility to acrylamide induced by changes in membrane fluidity.

The results of Gomez-Fernández *et al.* (1985) support the idea that membrane fluidity modulates the structure of the protein and that acrylamide may diffuse more easily through proteins when they are surrounded by a fluid phospholipid bilayer than when the surrounding lipid is in a gel state.

In line with these findings are the results of Almeida *et al.* (1984), who found that the Ca/ATP coupling ratio, i.e., the efficiency of the Ca^{2+} transport by the ATPase is significantly reduced by a decrease in membrane fluidity upon incorporation of cholesterol into sarcoplasmic reticulum membranes. The decrease in membrane fluidity is measured in this case by the use of the fluorescent probe dipyrenylpropane, a hydrophobic molecule that partitions into the lipid membrane.

ACKNOWLEDGMENTS. The work in the laboratory of the authors has been supported by Conselho Nacional de Desenvolvimento Científico e Tecnológico and Financiadora de Estudos e Projetos.

REFERENCES

Almeida, L. M., Vaz, W. L. C., Zachariasse, K. A., and Madeira, V. M. C., 1984, *Biochemistry* **23**:4714–4720.
Deamer, D., and Baskin, R. 1969, *J. Cell Biol.* **42**:296–307.
Dean, W. L., and Tanford, C., 1978, *Biochemistry* **17**:1683–1690.

de Meis, L., 1980, Transport and energy transduction in the sarcoplasmic reticulum, in: *Membrane Structure and Function* (E. Bittar, ed.), Wiley, New York.

Dupont, Y., 1976, *Biochem. Biophys. Res. Commun.* 71:544-550.

Dupont, Y., Bennett, N., Pougeois, R., and Lacapere, J., 1985a, in: *Structure and Function of Sarcoplasmic Reticulum* (S. Fleischer and Y. Tonomura, eds.), pp. 225-248, Academic Press, New York.

Dupont, Y., Harrisson, S., and Hasselbach, W., 1973, *Nature* 244:555-558.

Dupont, Y., and Leigh, J. B., 1978, *Nature* 273:396-398.

Dupont, Y., Pougeois, R., Ronjat, M., and Verjovski-Almeida, S., 1985b, *J. Biol. Chem.* 260:7241-7249.

Ebashi, S., and Lipman, F., 1962, *J. Cell. Biol.* 14:389-400.

Eftink, M. R., and Ghiron, C. A., 1981, *Anal. Biochem.* 114:199-227.

Esfahami, M., and Devlin, T. M., 1982, *J. Biol. Chem.* 257:9919-9921.

Guillain, F., Champeil, P., Lacapere, J. J., and Gingold, M. P., 1984, *Curr. Top. Cell. Regul.* 24:397-407.

Gomez-Fernández, J. C., Baena, M. D., Teruel, J. A., Villalain, J., and Vidal, C. J., 1985, *J. Biol. Chem.* 260:7168-7170.

Gratton, E., Jameson, D. M., and Hall, R. D., 1984, *Ann. Rev. Biophys. Bioeng.* 13:105-124.

Grubmeyer, C., and Penefsky, H. S., 1981, *J. Biol. Chem.* 256:3718-3727.

Hasselbach, W., and Makinose, M., 1961, *Biochem. Z.* 333:518-528.

Hasselbach, W., and Elfvin, L., 1967, *J. Ultrastructure Res.* 17:598-622.

Herbette, L., Marquardt, J., Scarpa, A., and Blasie, J., 1977, *Biophys. J.* 20:245-272.

Heremans, K. A. H., 1982, *Ann. Rev. Biophys. Bioeng.* 2:1-21.

Hahn, L-H. E., and Hammes, G. G., 1978, *Biochemistry* 17:2423-2429.

Hiratsuka, T., and Uchida, K., 1973, *Biochim. Biophys. Acta* 320:635-647.

Hudson, E. N., and Weber, G., 1973, *Biochemistry* 12:4154-4161.

Ikemoto, N., 1982, *Ann. Rev. Physiol.* 44:297-317.

Inesi, G., 1985, *Ann. Rev. Physiol.* 47:573-601.

Knopp, J., and Weber, G., 1969, *J. Biol. Chem.* 244:6309-6315.

Kurtenbach, E., and Verjovski-Almeida, S., 1985, *J. Biol. Chem.* 260:9636-9641.

Lacapere, J., Gingold, M., Champeil, P., and Guillain, F., 1981, *J. Biol. Chem.* 256:2302-2306.

Lakowicz, J. R., and Weber, G., 1973, *Biochemistry* 12:4171-4179.

Lehrer, S. S., 1971, *Biochemistry* 10:3254-3263.

Lentz, B. R., 1979, *Biophys. J.* 25:489-494.

London, E., and Feigenson, G., 1981, *Biochemistry* 20:1939-1948.

Lüdi, H., and Hasselbach, W., 1983, *Biochem. Biophys. Acta* 732:479-482.

Ludi, H., Rauch, B., and Hasselbach, W., 1982, *Z. Naturforsch.* 37c:299-307.

MacLennan, D. H., Brandl, C. J., Korczak, B., and Green, N. M., 1985, *Nature* 316:696-700.

Moczydlowski, E. G., and Fortes, P. A. G., 1981, *J. Biol. Chem.* 256:2357-2366.

Murphy, A., 1978, *J. Biol. Chem.* 253:385-389.

Paladini, A. A., Jr., and Weber, G., 1981, *Biochemistry* 20:2587-2593.

Perrin, F., 1926, *J. Phys. Rad.* 1:390.

Scott, T. L., 1985, *J. Biol. Chem.* 260:14421-14423.

Shinitzky, M., Dianoux, A. C., Gitler, C., and Weber, G., 1971, *Biochemistry* 10:2106-2113.

Shinitzky, M., and Rivnay, B., 1977, *Biochemistry* 16:982-986.

Silva, J. L., and Verjovski-Almeida, S., 1983, *Biochemistry* 22:707-716.

Silva, J. L., and Verjovski-Almeida, S., 1985, *J. Biol. Chem.* 260:4764-4769.

Spencer, R. D., and Weber, G., 1970, *J. Chem. Phys.* 52:1654-1663.

Teale, F. W. J., and Weber, G., 1957, *Biochem. J.* 65:476-482.

Vanderkooi, J. M., Ierokomas, A., Nakamura, H., and Martonosi, A., 1977, *Biochemistry* 16:1262-1267.

Vaz, W. L., Kaufmann, K., and Nicksch, A., 1977, *Anal. Biochem.* **83**:385-393.

Verjovski-Almeida, S., 1981, *J. Biol. Chem.* **256**:2662-2668.

Verjovski-Almeida, S., Kurtenbach, E., Amorim, A. F., and Weber, G., 1986, *J. Biol. Chem.* **261**:9872-9878.

Verjovski-Almeida, S., Kurzmack, M., and Inesi, G., 1978, *Biochemistry* **17**:5006-5013.

Verjovski-Almeida, S., and Silva, J. L., 1981, *J. Biol. Chem.* **256**:2940-2944.

Watanabe, T., and Inesi, G., 1982, *J. Biol. Chem.* **251**:11510-11516.

Weber, G., 1952, *Biochem. J.* **51**:145-155.

Weber, G., and Teale, F. W. J., 1957, *Trans. Faraday Soc.* **53**:646-655.

Yamamoto, T., Yantorno, R. E., and Tonomura, Y., 1984, *J. Biochem.* (*Tokyo*) **95**:1783-1791.

Yantorno, R. E., Yamamoto, T., and Tonomura, Y., 1983, *J. Biochem.* (*Tokyo*) **94**:1137-1145.

Yguerabide, J., Epstein, H., and Stryer, L., 1970, *J. Mol. Biol.* **51**:573-590.

Chapter 7

Lipid–Protein Interactions in the Function of the Na$^+$ and H$^+$ Pumps
Role of Sulfatide

F. Zambrano and M. Rojas

1. INTRODUCTION

The fluid mosaic model for membrane architecture proposed by Singer and Nicolson states that the integral proteins of membranes are globular amphypathic molecules that are either partially or totally inserted in the lipid bilayer (Singer, 1972). This model continues to have considerable explanatory and predictive value in the analysis of membrane structure and function. The fluid mosaic model depicts the membrane as a two-dimensional solution in which the lipid bilayer acts as a fluid solvent for the globular integral proteins. As a consequence of their mobility in the lipid bilayer, the membrane proteins are able to perform the varied processes of catalysis and transport carried out by biological membranes.

Most of the membrane components (oligosaccharides, lipids, and proteins) are asymmetrically oriented. In particular, integral membrane proteins are asymmetrically positioned across the membrane and their orientation is predominantly, or exclusively, in one or the other direction perpendicular to the plane of the bilayer (Singer, 1976). Membrane-bound transport enzymes that are asymmetrically oriented express their catalytic activities at only one surface of the membrane. Transport through the membrane by

F. Zambrano and M. Rojas • Departamento de Biología, Facultad de Ciencias, Universidad de Chile, Santiago, Chile.

these proteins is thus the result of both their anisotropic arrangement and the vectorial nature of enzyme catalysis under a given set of conditions (Nicolson and Singer, 1974).

The oligosaccharide chains of membrane glycoproteins of animal cells are exclusively exposed at the exterior surface of the plasma membrane or at the corresponding surface in intracellular membranes. Asymmetrical distribution is also found in the lipid components of the membrane. Phospholipids and glycolipids of most membranes are asymmetrically distributed in the two halves of the lipid bilayer (Verklaij et al., 1973; Tsai and Lenard, 1975; Nicolson, 1976). As long as there is an asymmetrical distribution of membrane-bound enzymes and lipids, a distinctive protein-lipid interaction could be expected at both sides of the membrane.

2. THE SODIUM PUMP

In 1957 Skou reported an adenosine triphosphatase in the microsomal fraction of the leg nerve of the shore crab that required Mg^{2+}, Na^+, and K^+ for optimal activity (Skou, 1957). He suggested that this enzyme might be involved in the coupled Na^+ and K^+ transport, which serves many important physiological functions, related to controlling Na^+ and K^+ gradients across membranes (Baker, 1972). Maintaining the gradients of Na^+ and K^+ in excitable tissues such as nerve and muscle provides the energy required to allow cellular excitability. It also affects transepithelial transport of Na^+ and, in some cases K^+, in such structures as the intestinal epithelium, the renal tubules, and various glandular structures (Edelman, 1974). Since Skou's discovery, the Na^+,K^+-ATPase has been extensively studied, and these studies have provided strong evidence that the Na^+,K^+-ATPase is the molecular structure responsible for affecting Na^+ and K^+ transport (Dahl and Hokin, 1974; Glynn and Karlish, 1975; Skou, 1975; Schuurmans Stekhoven and Bonting, 1981).

Findings of studies carried out on highly purified Na^+,K^+-ATPase preparation from a variety of tissues point out that the functional unit of Na^+ pump enzyme consists at least of two different subunits: a catalytic (α) subunit of about 100,000 mol.wt and a glycoprotein (β) subunit of about 50,000 mol.wt. The catalytic subunit contains both the site of phosphorylation and the cardiac glycoside binding site. The large subunit appears to span the membrane (Jørgensen, 1974; Perrone et al., 1975). The enzyme occurs in the membrane as a dimer, $\alpha_2\beta_2$ mol.wt about 250,000 (Kyte, 1975). At optimum conditions only about one half of the large subunits are phosphorylated (Lazdunsky, 1972). These observations lend some support to the proposal of a "flip-flop" mechanism for the pumping of cations across the membrane (Stein et al., 1973). As each subunit is involved in the overall

Na^+,K^+-ATPase reaction, the interaction between them should not be restricted by the lipid bilayer (Jørgensen, 1980; Schuurmans Stekhoven and Bonting, 1981).

3. LIPID REQUIREMENTS OF THE SODIUM PUMP: ROLE OF SULFATIDES

Because of its membrane localization, the Na^+,K^+-ATPase needs a lipidic microenvironment to perform its catalytic activity (Bonting and Caravaggio, 1963). Thus, the fluidity of the hydrocarbon chains present in membranes, together with other factors that modulate lipid–protein interactions, have an effect on ion translocation. Treatment of the Na^+,K^+-ATPase with detergents, phospholipases, or organic solvents leads to the inactivation of the enzyme. Thus, a clear correlation between the amount of lipid removed or hydrolyzed and the loss of enzyme activity can be established. Also, numerous studies indicate that phospholipid-depleted membrane enzymes can be reactivated by the sole addition of phospholipid (Dahl and Hokin, 1974). Although acidic phospholipids have been proposed as activators (Roloefson and Deenen, 1973), it has also been claimed that any phospholipid (acidic or neutral) could fulfill the specific lipid requirement for the activity of the sodium enzyme (De Pont et al., 1978). In spite of many attempts, it has been impossible to establish that reactivation of the inactivated enzyme requires a specific kind of phospholipid (De Pont et al., 1978). Thus, these findings suggest that factors other than phospholipids are involved in enzyme function.

Inasmuch as there is contradictory evidence about a specific phospholipid requirement for the enzyme involved in Na^+ and K^+ transport, we have focused our interest in Karlsson's findings suggesting that sulfatide, a glycolipid structural component of some plasma membranes, may play a function in the Na^+,K^+-ATPase activity (Karlsson, 1976). Karlsson has shown a high content of sulfatide in tissues specialized in active Na^+ ion transport like salt gland, electric organ, kidney, and nerve cells (Karlsson, 1976). Moreover, he showed that salt load in domestic duck produces an activation and hypertrophy of the salt gland with a concomitant induction of Na^+,K^+-ATPase activity and sulfatide content (Karlsson et al., 1971). Karlsson concluded that sulfatide is unique among membrane lipids in being specifically required for enzyme function. Based on these findings, Karlsson postulated a cofactor site model for sulfatide in Na^+,K^+-ATPase catalyzed reaction. In this model he suggests that the glycolipid component is essential for K^+ translocation, and that it functions as a K^+ site with its galactose-3-sulfate group on the outside face of the plasma membrane (Karlsson, 1976). However, to determine whether sulfatide has a real function it has to be shown that: (i) sulfatide is a structural component of membranes in which K^+ transport takes place; (ii) sulfatide is located in the outer half of plasma membrane bilayer; (iii) its removal or hydrolysis inactivates the Na^+,K^+-ATPase activity in proportion to the amount of sulfatide released or hydrolyzed, and (iv) that the enzyme inactivated by sulfatide removal can be reactivated by the sole addition of sulfatide. These ideas have stimulated further studies in our laboratory about the potential role of sulfatide on the Na^+,K^+-ATPase activity. Moreover, analysis of the lipid content (F. Zambrano, L. E. Hokin, and B. Fleischer, unpublished results) of homogenates of rectal gland of spiny dogfish and of electric organ of the electric eel, as well as of highly purified enzyme preparations isolated by Hokin et al. (Perrone et al., 1975), showed that sulfatide was present in large amounts in the purified ATPase preparation as compared to the initial homogenates (Table I).

It has been reported that the isolated Golgi apparatus from rat kidney is the subcellular locus for 3-phosphoadenosine 5-phosphosulfate: cerebroside sulfotransferase, the enzyme that catalyzes the conversion of cerebroside to sulfatide (Fleischer and Zambrano, 1974). It has also been shown that sulfatide occurs in all the cell fractions of kidney but is most abundant in the Golgi apparatus and plasma membrane (Zambrano et al., 1975). The fact that sulfatide formation takes place in the Golgi apparatus, and that it is a structural component of plasma membrane, is corroborated by following the rate of incorporation of radioactive sulfate into sulfatide of Golgi apparatus and plasma membrane of kidney cells (Fig. 1). The data are compatible with the idea that sulfatide is formed in vivo first in the Golgi apparatus and is then transferred by some as yet undefined mechanism to the plasma membrane (Fleischer et al., 1974).

Table I.   Sulfatide Content of Purified Na^+,K^+-ATPase Isolated from Two Different Sources[a]

	Rectal gland of spiny dogfish		Electric organ of eel	
	Homogenate	ATPase	Homogenate	ATPase
mg sulfatide/mg protein	0.062	0.182	0.032	0.162
Percent of total lipid	9.4	23.5	8.7	21.5
mg phosphatidylserine/mg protein	—	0.077	—	0.070

[a] Lipid was extracted from tissue homogenates and from the purified enzyme preparations using chloroform:methanol, 2:1 by volume. Nonlipid contaminants were removed by chromatography on Sephadex G-25. Total lipid was fractionated into neutral lipid (NL), phospholipid (PL), and glycolipid (GL) by column chromatography using silic acid. The amount of sulfatide was determined in the GL fraction by the method of Kean. The amount of phosphatidylserine was determined after two-dimensional thin-layer chromatography of the PL fraction, by the method of Chen et al. (Zambrano, Hokin, and Fleischer, unpublished results.)

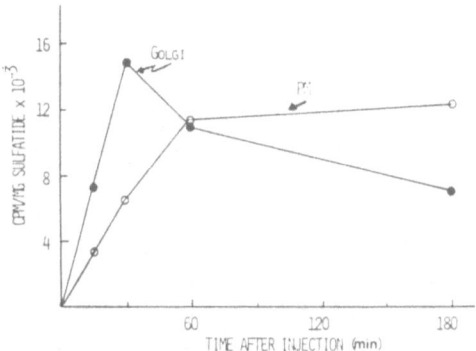

Figure 1. Appearance of radioactive sulfatide in isolated cell fraction from kidney after i.v. injection of [³⁵S]-sulfate, into rats. The dose was 0.5 mCi per rat. Golgi-rich and plasma membrane-rich fractions were isolated from the same sample of kidney. (From Fleischer *et al.*, 1974.)

The results obtained in kidney cell fractions strongly suggest that sulfatides are natural components of membranes in which Na⁺ and K⁺ transport occurs (Table II). These results suggest that these might have a role in the function of the ouabain-sensitive Na⁺,K⁺-ATPase enzyme. Following the idea that sulfatide might be a cofactor in the sodium pump, we examined the relationship between sodium transport Na⁺,K⁺-ATPase activity and sulfatide content in skin of different stages of larval development of the chilean frog *Calyptocephalella caudiverbera* (González *et al.*, 1979). It is known that during larval development the site of active Na⁺ and K⁺ transport changes its location from the abdominal skin (Alvarado and Moody, 1970) to the gill. The skin, in the advanced stages of development becomes the site where the Na⁺ translocation takes place; meanwhile the gill, in the same stages, reduces its physiological function to a very low level.

Measurement of the sulfatide content in the glycolipid fraction, and of the ouabain sensitive Na⁺,K⁺-ATPase activity in total homogenates of

Table II. Lipid Content of Rat Kidney Subcellular Fractions[a]

	Mitochondria	Rough endoplasmic reticule	Golgi apparatus	Plasma membrane
NL	0.040	0.116	0.162	0.216
PL	0.215	0.395	0.537	0.813
GL	0.012	0.020	0.147	0.071
Sulfatide	0.016	0.022	0.120	0.055
Total lipid	0.267	0.531	0.848	1.10

[a] All values are expressed as mg/mg protein. Lipid was extracted from the fractions and fractionated as indicated in Table I. The amount of lipid in the NL, PL, and total lipid fraction was determined by drying and weighing an aliquot of the fraction. The amount of glycolipid in the GL fraction was determined by analysis for sphingosine. Sulfatide was determined in the GL fraction by the method of Kean. (From Zambrano *et al.*, 1975.)

abdominal skin during the larval development, indicated that the increase with time, reflected in the slopes of the ATPase activity curve between stages 36 and 48, is 0.178 ± 0.0001, while for the sulfatide content it is 0.032 ± 0.0001 (Fig. 2). The ratio between the slopes was 1.87, which indicates that as a function of development time, sulfatide content increases almost twice as fast as Na^+,K^+-ATPase activity. On the other hand, ATPase activity greatly increases in the larval skin during this period; the activity at stage 44 is sevenfold greater than at stage 36. Furthermore, the values of sodium flux obtained in the abdominal skin from three different larval stages of development (Table III) indicates that in younger stages of development (stages 30–35), the skin of these larvae presents a sodium flux that remains almost unchanged by amiloride addition. In stages 36–39 the sodium flux is more sensitive to amiloride, while in more advanced stages the inhibition by

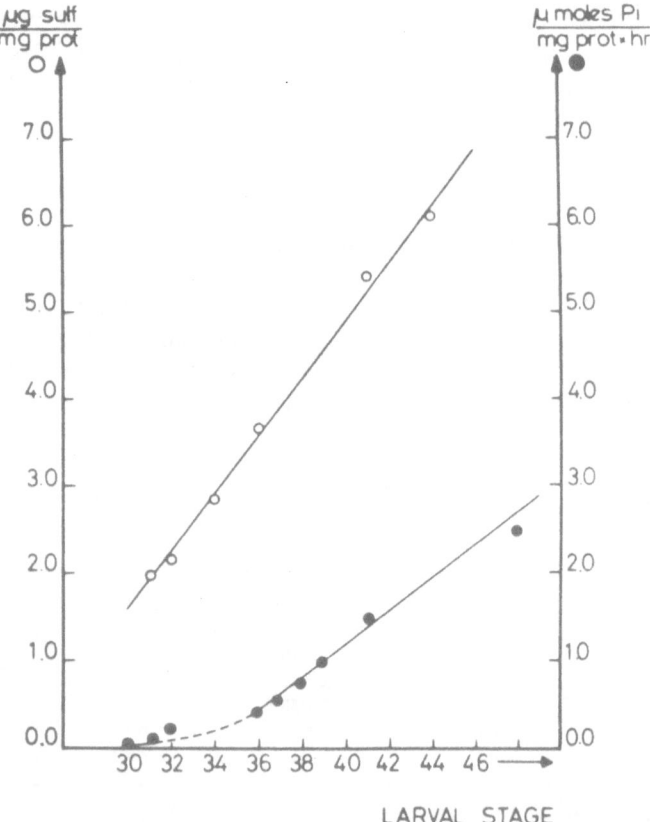

Figure 2. Sulfatide content (open symbols) and Na^+,K^+-ATPase (ouabain-sensitive) activity (solid symbols) in total homgenate of larval skin obtained from *Calyptocephalella caudiverbera* at different stages of development. (From González *et al.*, 1979.)

Table III. Relationship between Net Sodium Flux and Larval Development[a]

Stage of Larval development	Sodium flux in Ringer solution (nmol/cm² hr)		Sodium flux in Ringer solution + amiloride (nmol/cm² hr)	
	$\bar{X}$	Range of values	$\bar{X}$	Range of values
30–35 (6)	0.90	0.55–1.37	0.79	0.55–1.37
36–39 (4)	7.54	5.48–8.91	1.83	1.37–2.74
40–44 (4)	220.00	80.00–300.00	9.25	4.00–14.00

[a] Sodium flux in abdominal skin of different stages of development. The measurements were performed at 20°C. Amiloride in a final concentration of 0.058 mM was used. (From González *et al.*, 1979.)

amiloride is about 96%. Comparing the last two cases (stages 36–39 with stages 40–44), the sodium flux through the abdominal skin is quite different, becoming 37-fold higher in the advanced stages than in the younger ones. So far the studies carried out strongly suggest that tissues rich in ouabain sensitive Na^+,K^+-ATPase activity associated with sodium ion transport present a high concentration of sulfatide and that the rise in sulfatide content precedes the rise in Na^+,K^+-ATPase (Fig. 2; Table III).

The phospholipid/protein ratio, which reflects the amount of membrane present, is not significantly different during the larval development in the gill; and in the skin, this ratio increases 1.6 times along with development (Table IV). The negatively charged phosphatidylserine has been postulated as a specific requirement for Na^+,K^+-ATPase activity (Roloefson and Deenen, 1973). It could therefore be expected to show significant concentration changes during larval stage development. Nevertheless, during larval development the amount of phosphatidylserine in both tissues is fairly constant. In addition, the total conversion of phosphatidylserine to phosphatidylethanolamine by specific enzymatic decarboxylation of a purified Na^+,K^+-ATPase from rabbit kidney outer medulla resulted in a loss of only 13% of the activity (De Pont *et al.*, 1978), suggesting that there is not a specific phosphatidylserine requirement for this enzyme.

On the other hand, the treatment of the microsomal fraction of abdominal skin of advanced stages of larval development with arylsulfatase A, a lysosomal enzyme involved in sulfatide hydrolysis giving cerebroside and sulfate as products, inhibited the ouabain sensitive Na^+,K^+-ATPase activity. A 36% breakdown of sulfatide inhibited 100% of the ouabain sensitive ATPase (Fig. 3). The enzyme activity in preparations treated with arylsulfatase A, free of other enzymes or inhibitors such as proteases or other hydrolases, can be recovered by the sole addition of sulfatide microdispersion (González *et al.*, 1979). The Na^+,K^+-ATPase activity thus recovered is still ouabain sensitive. These results seem to indicate that

Table IV. Sulfatide and Phospholipid Content of Skin and Gill of Different Larval Stages of *Calyptocephalella caudiverbera* (100,000 × g pellet)[a]

	Skin			Gill		
	30–35	36–39	40–44	30–35	36–39	40–44
Larval stage						
Ouabain-sensitive Na⁺,K⁺-ATPase activity (μmol Pi/mg hr)	N.D.	1.40	2.15	0.87	3.29	N.D.
(μg P/mg protein)	11.09	11.07	18.72	14.80	13.83	13.86
(μg P of phospholipids/mg protein	2.16	2.08	3.44	3.94	3.91	3.77
μg sulfatide/mg protein	13.94	42.60	55.44	25.56	10.94	6.17
μg of phospholipids/mg protein	51.84	49.62	82.56	70.56	93.84	90.48
μg sulfatide/μg phospholipids	0.27	0.85	0.67	0.36	0.12	0.07
Percent phosphatidylserine total phospholipids	8.15	8.42	11.11	10.83	7.02	9.97

[a] A factor of 24 was used to calculate the amount of phospholipids based on a mean molecular weight of 744 (From González *et al.*, 1979).

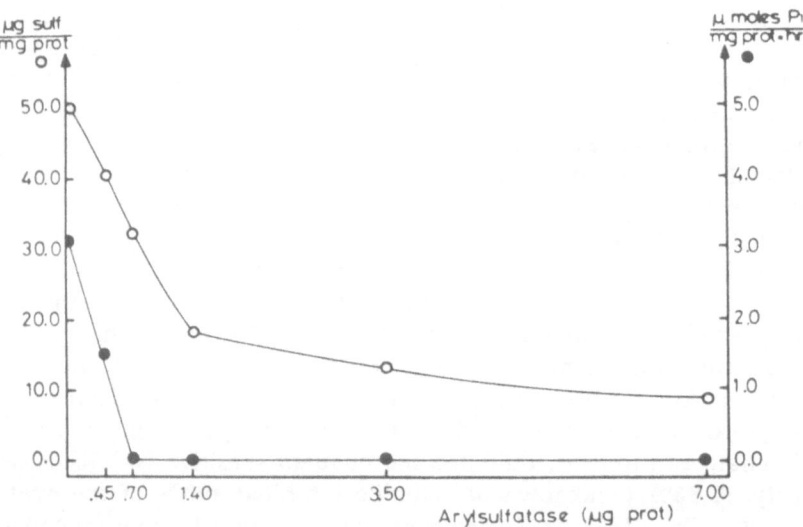

Figure 3. Inhibition of Na⁺,K⁺-ATPase activity (solid symbols) produced by the specific hydrolysis of sulfatide. Microsomal fraction of skin was incubated for 10 min with different amounts of arylsulfatase and incubated an additional 10 min with ATP. Half the final volume was used to measure the Pi release; the other half was used to determine sulfatide content (open symbols). (From González *et al.*, 1979.)

sulfatide may play a role in the ATPase activity and, as a consequence, in the active transport of Na^+ and K^+.

Although the hydrolysis of ATP is a step in Na^+ and K^+ translocation, the mentioned studies on lipid requirement for ATP hydrolysis do not prove whether sulfatide is essential for the translocation process. To understand the physiological role of sulfatide in cell membrane function and to define more precisely the function of sulfatide in Na^+ and K^+ transport, we studied the sodium efflux in ^{22}Na-loaded red blood cells in the presence of arylsulfatase (Zambrano et al., 1981), knowing that sulfatide is also a structural component of this membrane (Hansson et al., 1978).

The results showed that sodium efflux, calculated from the slopes of the ^{22}Na loss, diminishes as the amount of arylsulfatase increases in Ringer's solution (Table V). The highest amount of arylsulfatase added reduced the active sodium efflux (defined as the ouabain-sensitive sodium efflux) to about 74%, while the reduction observed in the potassium-free medium was of the order of 72%. It was also found that with 1.20 units of arylsulfatase (a unit is defined as the amount of enzyme that hydrolyzes 1 nmol of p-nitrocatechol sulfate per min at 37°C, pH 5.4), the Na^+ efflux from red cells in K^+-free Ringer's solution and Ringer's solution with or without ouabain were not significantly different. On the other hand, the sodium effluxes in K^+-free medium with or without the highest amount of arylsulfatase were very similar. Moreover, the inhibition of sodium efflux induced by the highest amount of arylsulfatase was diminished in the presence of an exogenous sulfatide microdispersion (Table VI). Thus, 10 μg of sulfatide in the medium decreases the inhibition level of 61% of the sodium efflux caused by the arylsulfatase to about 20%, and with 20 μg of sulfatide the sodium efflux is 21% higher than the efflux in Ringer's solution alone, even with the arylsulfatase present.

A straight correlation between sulfatide degradation and Na^+,K^+-ATPase activity reduction was observed in the fragmented membrane isolated from the same loaded red cell from which the ^{22}Na loss curves were obtained. The data showed that the highest sulfatide breakdown level caused an inhibition of 100% of the ouabain-sensitive Na^+,K^+-ATPase, leaving unaffected the ouabain-insensitive activity (Table VII).

So far, the data demonstrate that membrane loaded red cells treated with arylsulfatase show sulfatide breakdown, that cells are not hemolyzed when sodium efflux is reduced by about one third, and that the addition of sulfatide is able to protect the cell from the action of arylsulfatase. These findings seem to indicate that sulfatide is located externally in the red blood cell membrane and that it is involved in the sodium pump. In the absence of external potassium a one-for-one sodium exchange and reduction of sodium efflux by about one third has been observed. Similar reduction is obtained in the presence of the highest amount of arylsulfatase used.

Table V. Effect of Arylsulfatase on Sodium Efflux in Red Blood Cells[a]

	Arylsulfatase (Units/ml)	Ringer's solution	Inhibition (%)	Ouabain (70 μM)	Inhibition (%)	K-Free medium	Inhibition (%)	100 mM K medium
Na efflux (mmoles/liter of cells/hr)	None	3.34	0.00	0.90	73.05	1.59	52.39	3.34
	0.30	2.91	12.87					
	0.60	2.20	34.13					
	1.20	1.45	56.59	0.85	74.55	1.53	54.19	1.61
Active Na Efflux (mmoles/liter of cells/hr)	None	2.44	17.62	0.00	100.00	0.69	71.72	2.44
	0.30	2.01	46.72					
	0.60	1.30						
	1.20	0.55	77.46	0.00	100.00	0.63	74.18	0.71

[a] Na efflux in red blood cells suspended in different media. The values were calculated from the slopes of ^{22}Na loss curves. (From Zambrano et al., 1981.)

Table VI. Effect of Sulfatide on Sodium Efflux in Red Blood Cells[a]

	Arylsulfatase (units/ml)	Sulfatide (μg/ml)	Ringer's solution	Inhibition (%)	Ouabain (70 μm)	Hemolysis range of values (%)
Na efflux (mmoles/liter of cells/hr)	None	None	3.34	0.00	0.90	N.D.
	None	2.5	3.31	0.90		0.00–1.60
	None	5.0	3.31	3.89		1.80–2.50
	1.20	5.0	1.31	60.78		N.D.
	None	10.0	3.22	3.59		5.40–7.10
	1.20	10.0	2.66	20.36		N.D.
	1.20	20.0	3.86	0.00		N.D.
Active Na efflux (mmoles/liter of cells/hr)	None	None	2.44	0.00	100.00	
	None	2.5	2.41	1.22		
	None	5.0	2.31	4.15		
	1.20	5.0	0.41	82.98		
	None	10.0	2.32	4.92		
	1.20	10.0	1.76	27.87		
	1.20	20.0	2.96	0.00		

[a] Na efflux in red blood cells in the presence of an exogenous microdispersion of sulfatide and/or arylsulfatase (from Zambrano et al., 1981).

Table VII.   Effect of Arylsulfatase on the Loaded Erythryocyte Membranes[a]

Arylsulfatase	Total ATPase	Ouabain insensitive ATPase	Percent sulfatide degradation after incubation		
			20 min	40 min	60 min
None	1.39	0.94	N.D.	N.D.	N.D.
0.30 unit	1.26	0.95	21.5	40.1	58.9
0.60 unit	1.17	0.95	27.2	46.6	69.2
1.20 unit	0.93	0.93	32.6	57.1	83.2
1.20 unit plus 20 μg sulfatide	1.67	0.96	N.D.	N.D.	N.D.

[a] The table illustrates the relationship between Na^+,K^--ATPase activity and sulfatide degradation in the membrane of loaded red cells suspended in medium with arylsulfatase (from Zambrano *et al.*, 1981).

Inasmuch as the sulfatide cofactor model postulates affinity between sulfatide and potassium ions (Karlsson, 1976), the affinity site for potassium on the external side of the membrane would be inactive when sulfatide is absent, and total inhibition of potassium influx would be expected. The sodium efflux findings in human red blood cells suggest that the sulfatide are involved in the sodium pump either in the specific K^- binding site or in the affinity site for K^+.

As the mentioned results suggest that sulfatide in general is required for the function of the sodium pump, and, in particular, is involved in the affinity site for K^+, we focused our interest in the partial reactions which describe the transport process, and especially in the step that requires K^+. The mechanism of this enzyme can be summarized briefly in the following partial reactions which describe the transport process (Albers, 1976).

$$E_1 + ATP + 3Na_i^+ \rightleftarrows E_1 - P(Na^+)_3 + ADP$$

$$E_1 - P(Na^+)_3 + Mg^{2+} \rightleftarrows E_2 - P + 3Na_0^+$$

$$E_2 - P + 2K_0^+ \rightleftarrows E_2(K^+)_2 + P_i$$

$$E_2(K^+)_2 \rightleftarrows E_1 + Mg^{2+} + 2K_i^-$$

The Na^+,K^+-ATPase exists in two forms, which are believed to have different conformations and thus different affinities for cations. E_1 is the form of the enzyme with the cation binding site oriented inward, while E_2 is the form of the enzyme with its cations binding sites oriented outward (Post *et al.*, 1965). The reaction begins with the enzyme being phosphorylated by ATP in the presence of Mg^{2+} and Na^+ at the inner surface of the cell membrane. The phosphoenzyme is then hydrolyzed in the presence of K^-. The binding site for K^+ and ouabain is on the outside of the cell membrane.

It is known that the Na^+,K^+-ATPase from a wide variety of sources catalyzes the Na^+,K^+ activated, ouabain-sensitive hydrolysis of p-nitrophenylphosphate. The close association of both enzyme activities suggests that the phosphatase is involved in the final step of the reaction catalyzed by the Na^+,K^+-ATPase, hydrolyzing the phosphate ester provided by the Na^+-dependent phosphorylation of the enzyme. If sulfatide is involved in the affinity site for K^+, this site would be missing when sulfatide is absent and inhibition of K^+-activated phosphatase activity would be expected, since K^+ would be unable to reach the inside of the cell, the K^+ site of the enzyme.

To test this hypothesis, we isolated a microsomal fraction rich in Na^+,K^+-ATPase from the outer medulla of pig kidney, and Mg^{2+},K^+ activated phosphatase activity was studied in this preparation treated with arylsulfatase (González and Zambrano, 1983). The data obtained showed that the phosphatase activity is reduced when sulfatide is hydrolyzed. With 62% of the sulfatide hydrolyzed, the phosphatase activity is reduced to 23% of the control. As the ouabain-insensitive activity is not affected, this represents an almost total (98%) inhibition of the ouabain-sensitive activity (Table VIII). The arylsulfatase treatment causes inactivation on the ouabain-sensitive phosphatase even in the presence of K^+ concentration as high as 100 mM (Table IX). The loss of the ouabain sensitive phosphatase activity of the arylsulfatase-treated microsomal fraction could be reversed by the sole addition of exogenous sulfatide. A maximum reactivation of almost 70% of the ouabain-sensitive phosphatase activity was obtained with the addition of 24 times the sulfatide content of intact microsomal preparation (Table X). The fact that the hydrolysis of sulfatide abolishes the dephosphorylation step and that this effect can be reversed by the addition of

Table VIII.   Effect of Arylsulfatase on Phosphatase Activity[a]

Arylsulfatase units	Ouabain (mM)	Specific activity (nmol p-nitrophenol/ min mg)	Sulfatide hydrolyzed (%)	Inhibition (%)	Inactivation (%)
None		498 ± 60			
1		357 ± 32	5.6 ± 0.8		28.4
2		248 ± 23	23.4 ± 2.1		50.3
4		115 ± 10	59.1 ± 4.9		77.0
8		114 ± 9	62.3 ± 5.7		77.1
None	1	112 ± 8		77.5	
4	1	112 ± 8		77.6	

[a] The table shows the effects of arylsulfatase on the ability of Na^+,K^+-ATPase preparation to hydrolyze p-nitrophenylphosphate (from González and Zambrano, 1983).

Table IX. Effect of Arylsulfatase on Phosphatase Activity in the Presence of Different K^+ Concentrations[a]

K^+ (mM)	Arylsulfatase units	Ouabain (mM)	Specific activity (nmol p-nitrophenol/ min mg)	Inactivation (%)	Inhibition (%)
0	None		114 ± 8		
0	None	1	112 ± 13		
0	4		115 ± 7		
8	None		498 ± 60		
8	None	1	112 ± 8		77.5
8	4		115 ± 10	77.0	
20	4		113 ± 9	77.3	
80	4		115 ± 5	77.0	
100	4		115 ± 6	77.0	
100	None	1	112 ± 5		77.5

[a] From González and Zambrano (1983).

Table X. Reactivation of K^+-Dependent Ouabain Sensitive Phosphatase Activity in Arylsulfatase-Treated Microsomal Fraction[a]

	Specific activity (nmol p-nitrophenol/ min mg)	Inactivation (%)	Inhibition by ouabain (%)	Reaction of ouabain sensitive K^+-phosphatase (%)
Untreated	393	99.0		
Treated	4 ± 2			
Treated plus sulfatide				
5.4	93 ± 12			23.8
10.8	223 ± 19			61.4
16.2	346 ± 29			88.1
21.6	351 ± 33			89.4
27.0	289 ± 37			73.6
26.6 + ouabain	11 ± 2		97.3	
Treated plus phospholipid				
40.0	17 ± 3			
80.0	20 ± 5			
160.0	42 ± 7			
320.0	68 ± 17			
320.0 + ouabain	70 ± 10			

[a] The different amounts of lipid (μg) used to reactivate correspond to 6, 12, 18, 30 and 2.5, 5, 10, and 20 times the sulfatide and phospholipid content of the microsomal preparation used, respectively (from González and Zambrano, 1983).

sulfatide seems to indicate that sulfatide plays a role in the affinity site for potassium in the outside of the membrane. K^+-activated phosphatase activity in the absence of ATP and Na^+ is stimulated by K^+ at the cytoplasmic surface (Drapeau and Blastein, 1980). Thus, when sulfatide is absent, inactivation of K^+-activated phosphatase activity would be expected as sulfatide has been proposed as essential for K^+ translocation.

Hence, these results, together with the demonstration that there is straight correlation between Na^+,K^+-ATPase activity, sulfatide content, and sodium flux in skin (González et al., 1979), and that the specific hydrolysis of sulfatide, localized externally in the red cell membrane, reduced the sodium efflux to the same extent as the efflux in the absence of external potassium (Zambrano et al., 1981), strongly suggest that sulfatide is involved in the specific affinity site for K^+. If this is the function of sulfatide, it would be expected that in an ATPase preparation treated with arylsulfatase the K^+-dependent dephosphorylation as well as the ouabain binding should be blocked.

Treatment of the partially purified Na^+,K^+-ATPase from outer medulla of fresh pig kidney with arylsulfatase indicated that in sharp contrast to the K^+-dependent dephosphorylation (Jedlicki and Zambrano, 1985), the steady-state level of Na^+-dependent phosphorylation remained unaffected. The phosphorylated intermediate is completely K^+ insensitive to K^+-induced dephosphorylation when 63% of the original sulfatide is hydrolyzed. The K^+ sensitive of the phosphoenzyme is lost after arylsulfatase treatment, and is similar to that obtained on untreated microsomal fraction in the presence of ouabain (Table XI). The inactivation produced by arylsulfatase treatment on K^+-dependent dephosphorylation was gradually

Table XI. Effect of Arylsulfatase on Phosphorylation of Na^+,K^--ATPase Preparation with $[\gamma^{-32}P]$-ATP[a]

Ionic medium	Arylsulfatase (units)	Ouabain (mM)	Steady state phosphorylation (pmoles/mg)
$Mg^{2+}Na^+$			$708 \pm 30\,(8)$
$Mg^{2+}Na^+K^+$			$174 \pm 15\,(6)$
$Mg^{2+}K^+$			$82 = 6\,(5)$
$Mg^{2+}Na^+K^+$		1	$737 \pm 39\,(3)$
$Mg^{2+}Na^+$	20		$750 \pm 52\,(4)$
$Mg^{2+}Na^+K^+$	20		$701 \pm 40\,(5)$

[a] The ATPase preparation, after incubation with or without arylsulfatase for 20 min at 37°C in K^+-free medium, was centrifuged at $100,000 \times g$ for 15 min. The pellet was resuspended and aliquots were incubated with $[\gamma^{-32}P]$-ATP in different ionic media. An aliquot of the pellet was extracted with chloroform : methanol (2 : 1 v/v). The amount of sulfatide was measured in the isolated glycolipid fraction. (From Jedlicki and Zambrano, 1985.)

abolished by increasing the sulfatide concentration, reaching about 100% protection in the presence of 30 times the endogenous sulfatide content (Table XII). The ability of Na^+,K^+-ATPase preparation to bind ouabain decreased after treatment with arylsulfatase. Twenty units of arylsulfatase, which hydrolyzed 63% of the sulfatide content, inactivated the ouabain binding by about 96% (Table XIII). As in K^+-dependent dephosphorylation, the inactivation produced by arylsulfatase on ouabain binding decreased when sulfatides were added together with arylsulfatase in the preincubation medium (Table XIV).

It is necessary to point out that the ability of the untreated Na^+,K^+-ATPase preparation to form Na^+-dependent and independent phosphoenzyme as well as to carry out K^+-dependent dephosphorylation, together with the ability to bind ouabain and its K^+-activated phosphatase activity, was unaffected by exogenous sulfatide addition. Properties of the Na^+,K^+-ATPase preparation lost by hydrolysis of sulfatide can be reverted by the sole addition of sulfatide, in contrast to phospholipids that are unable to reactivate these activities (González and Zambrano, 1983; Jedlicki and Zambrano, 1985).

The fact that the tested Na^+-dependent ADP-ATP exchange activity remained unaffected after arylsulfatase treatment of the Na^+,K^+-ATPase preparation (Table XV) indicates that sulfatide is involved in subsequent partial reaction steps, perhaps in the conformation transition of the K-insensitive phosphointermediate, E_1P, to the K^--sensitive intermediate, E_2P. Such an interpretation would also explain the data on ouabain binding if it is assumed that E_1P has a low affinity for ouabain as compared to E_2P. However, if the hydrolysis of sulfatide stabilizes the protein in the E_1P form by blocking the passage to E_2P, an increase of the E_1P form would be expected with a concomitant increase of ADP-ATP exchange activity since

Table XII. Effect of Arylsulfatase on Phosphorylation of Na^+,K^+-ATPase Preparation in the Presence of Microdispersed sulfatide[a]

Ionic medium	Arylsulfatase (units)	Sulfatide (μg)	Steady state phosphorylation (pmoles/mg)
$Mg^{2+}Na^+$			$626 \pm 32\,(8)$
$Mg^{2+}Na^+K^+$			$92 \pm 8\,(6)$
$Mg^{2+}Na^+K^+$	20		$620 \pm 39\,(5)$
$Mg^{2+}Na^+K^+$	20	92.8	$209 \pm 10\,(4)$
$Mg^{2+}Na^+K^+$	20	185.6	$131 \pm 20\,(5)$
$Mg^{2+}Na^+K^+$	20	278.4	$92 \pm 6\,(3)$

[a] Sulfatide added correspond to 10, 20, and 30 times the content in the ATPase preparation used ($46.4\,\mu$g per mg of protein) (from Jedlicki and Zambrano, 1985).

Table XIII. Effect of Arylsulfatase on [^{3}H]-Ouabain Binding to Na$^+$,K$^+$-ATPase Preparation[a]

Arylsulfatase (units)	Ouabain binding (pmoles/mg)	Binding (%)	Inactivation (%)
None	504 ± 64 (8)	100	
5	76 ± 34 (6)	15	85
10	46 ± 24 (4)	9	91
20	19 ± 12 (5)	4	96

[a] The ATPase (200 μg) preparation, after incubation with or without arylsulfatase for 20 min at 37°C in the appropriate ionic medium, was centrifuged at 100,000 × g for 15 min. Pellets were resuspended and incubated in ionic medium with ouabain and 0.1 μCi [^{3}H]-ouabain for 30 min at 37°C. (From Jedlicki and Zambrano, 1985.)

this activity is proportional to the level of E_1P. However, no change in ADP–ATP exchange was observed. Alternatively, sulfatide might not be involved in the E_1P-E_2P conformational transition but in the configuration of the high-affinity binding site for K$^+$ of the E_2P. In the absence of sulfatide, the reaction with K$^+$ ions attracted to the neighborhood of the ATPase translocation site would be inhibited.

According to the sulfatide cofactor model (Karlsson, 1976) the arylsulfatase-modified E_1P intermediate would remain phosphorylated because the K$^+$ needed to start dephosphorylation process would not interact with E_2P, resulting in a situation similar to that provoked by ouabain. It is also possible that sulfatides play a role in the transition E_2PK-E_2K + P.

Thus, the present results indicate that sulfatides are involved in a partial reaction step of the sodium pump enzyme subsequent to the formation of E_1P. However, to envisage the particular role of sulfatide in the sodium pump enzyme further studies, such as potassium influx, potassium binding, and interconversion of the phosphoenzyme form, will be needed.

Table XIV. Effect of Arylsulfatase on [^{3}H]-Ouabain Binding to Na$^+$,K$^+$-ATPase Preparation in the Presence of Microdispersed Sulfatide[a]

Arylsulfatase (units)	Sulfatide (μg)	Ouabain binding (pmoles/mg)	Inactivation (%)
None		504 ± 64 (8)	
None	185.6	471 ± 6 (5)	6
5		76 ± 34 (6)	85
5	92.8	192 ± 39 (3)	62
5	185.6	238 ± 12 (3)	53

[a] Sulfatide added correspond to 10 and 20 times the sulfatide content of 200 μg of protein of ATPase preparation (from Jedlicki and Zambrano, 1985).

Table XV.   Effect of Arylsulfatase on Na^+-Dependent ADP–ATP Exchange Activity[a]

Ionic medium	Arylsulfatase (units)	Ouabain (mM)	Rate of incorporation (nmoles/mg/min)
Mg^{2+}		0.1	24 ± 4
Mg^{2+}	20	0.1	26 ± 5
$Mg^{2+}Na^+$			158 ± 11
$Mg^{2+}Na^+$	20		163 ± 8
$Mg^{2+}Na^+$	20	0.1	22 ± 1

[a] The ATPase preparation (200 μg), after incubation with or without arylsulfatase for 20 min at 37°C in the appropriate ionic medium, was centrifuged as indicated in Table XIII. Pellets were resuspended and aliquots incubated in ionic medium with unlabeled ATP and [^{14}C]-ADP for 20 min at 30°C. (From Jedlicki and Zambrano, 1985.)

4. ROLE OF SULFATIDES IN THE PROTON PUMP

It is possible that sulfatide is generally involved in affinity sites for K^+, in membrane enzymes activities in which K^+ translocation takes place. The ATP system of the apical membrane of the oxyntic or parietal cells is involved in a process of acid secretion activated by K^+. If the role of sulfatide in the activation of K^+ binding site is general to membrane-associated transport enzymes, then sulfatide should play a role in the proton pump system. To test the possible role of sulfatide in the mechanism of H^+ and K^+ transport we analyzed the sulfatide content of a vesicular membrane fraction from rabbit gastric mucose. In addition, we investigated the activities of H^+,K^+-ATPase and K^+-activated phosphatase as well as the H^+ pumping after treatment with arylsulfatase (Zambrano and Rojas, 1986).

Our preliminary studies show that the isolated vesicular membrane fractions have activities of both K^+-stimulated, Mg^{2+}-dependent ATPase and phosphatase higher than that tested in homogenate and/or microsomal fraction, with a contaminant increase in the capacity of H^+ pumping (Table XVI). In addition to the increased enzyme activities, the fraction is enriched in cholesterol, phospholipid, and sulfatide. The phospholipid and cholesterol contents, which constituted about 86% and 30% of total lipid, are similar to lipids distributions of other vesicular membrane fractions reported previously (Sen and Ray, 1979). The sulfatide content in the vesicles is about three times that of the homogenate and accounts for 10% of total lipid composition (Table XVII). The enzyme activities of Mg^{2+} ATPase and K^+-stimulated phosphatase of the vesicular preparation treated with arylsulfatase remained unaffected, in sharp contrast with the K^+-activated ATPase activity. The gradual specific hydrolysis of sulfatide abolished in the same way the activity of H^+,K^+-ATPase activity, and when 75% of sulfatide is

Table XVI.   Enzymatic Characterization of the Vesicular Membrane Fraction[a]

		K⁺-stimulated, ouabain insensitive enzyme activities		
	Mg^{2+}-ATPase	H$^+$,K$^+$-ATPase	Phosphatase	H$^+$ pumping
Homogenate	9 ± 2	3 ± 1	2 ± 1	N.D.
Microsomal fraction	20 ± 5	20 ± 2	18 ± 4	38 ± 8
Vesicular membrane fraction	12 ± 3	35 ± 4	33 ± 9	118 ± 29

[a] Enzymatic activities are expressed as μmol of Pi or p-nitrophenol released per hour and per milligram of protein. The H$^+$ pumping is expressed as μmol of H$^+$ per milligram of protein measured by titration with HCl. The values represent the mean ± SD of five experiments (Zambrano and Rojas, 1986).

hydrolyzed almost total inactivation of this particular enzyme activity is observed (Table XVIII). The addition of sulfatide to vesicular membrane fraction pretreated with arylsulfatase restored the K$^+$-stimulated ATPase activity (Table XIX). A maximum 46% of reactivation was reached by incubation with 7.5 times the endogeneous sulfatide content. The highest addition of sulfatide was unable to improve the reactivation. At the concentration of sulfatide added that was able to reactivate the enzyme, partial amounts of sulfatide were bound to the vesicular membrane fraction and no exchange with the phospholipid component was detected. On the other hand, different amounts of microdispersed phospholipids were unable to restore the lost activity of H$^+$,K$^+$-ATPase produced by arylsulfatase treatment (data not shown).

In general, the site of activation by K$^+$ and ATP of this enzyme is reversed compared with Na$^+$,K$^+$-ATPase: ATP activates at the intravesicular side, whereas K$^+$ activates at the extravesicular side and is transported outward in exchange for protons. Thus, the meanings of E$_1$ and E$_2$ are also reversed: in the E$_1$ conformation the cation binding sites are outwardly directed, whereas in the E$_2$ conformation they are inwardly directed. The associated K$^+$ activated phosphatase activity is catalyzed by

Table XVII.   Lipid Composition of the Vesicular Membrane Fraction[a]

	Phospholipid	Cholesterol	Sulfatide
		(μg per mg of protein)	
Homogenate	254 ± 38	52 ± 6	51 ± 8
Microsomal fraction	277 ± 36	152 ± 18	80 ± 9
Vesicular membrane fraction	646 ± 71	465 ± 47	183 ± 31

[a] Lipid from homogenate and fractions were extracted and fractionated as indicated in Table 1. The values represent the mean ± SD of 10 experiments (Zambrano and Rojas, 1986).

Table XVIII. Effect of Arylsulfatase on Enzymatic Activities of the Vesicular Membrane Fraction[a]

Arylsulfatase added (μg)	Sulfatide hydrolyzed (%)	Mg^{2+}-ATPase	Phosphatase	H^+,K^+-ATPase	Inactivation (%)
None		12 ± 3	33 ± 9	35 ± 4	
20	6.2	12 ± 4	34 ± 12	24 ± 2	31.4
40	30.7	12 ± 6	35 ± 13	15 ± 2	57.1
80	75.2	12 ± 3	35 ± 16	2 ± 1	94.3

[a] Aliquots (1.5 mg of vesicular membrane protein) were treated for 10 min with different amounts of arylsulfatase. Aliquots of the treated membrane were used to determine the enzymes activities, and the remaining volume was used to isolate and determine sulfatide. Enzymatic activities are expressed as in Table XVI. The values represent the mean ± SD of 5 preparations (Zambrano and Rojas, 1986).

a K^+ site located on the outer side of the vesicles. In the acid secretion process, potassium enter the vesicles as KCl and then exchanges by protons in the ATP-driven process leaving HCl inside the vesicles to be expelled outside the cell (Schuurmans Stekhoven and Bonting, 1981).

So far, our preliminary results obtained in vesicular membrane fraction from mucosal gastric microsomes of rabbit seem to indicate that sulfatide also may play a function in the high affinity binding site for K^+ of the enzyme involved in the acid secretion process (Schuurmans Stekhoven and

Table XIX. Reactivation of H^+,K^+-ATPase of Arylsulfatase-Treated Vesicular Membrane Fraction by Added Microdispersed Sulfatide[a]

	Sulfatide added (mg/mg protein)	Enzyme activity	Sulfatide-bound (μg/mg protein)	Phospholipid content (μg/mg protein)
Untreated	None	35 ± 4	183 ± 31	646 ± 97
Treated	None	2 ± 1	45 ± 10	650 ± 110
	0.46	7 ± 3	253 ± 48	643 ± 93
	0.92	13 ± 2	291 ± 53	638 ± 112
	1.38	29 ± 4	313 ± 59	626 ± 115
	1.84	18 ± 4	334 ± 58	605 ± 95
	2.30	9 ± 2	354 ± 60	571 ± 91
	3.68	6 ± 2	379 ± 67	507 ± 100

[a] Vascular membrane fraction (1.5 mg of protein) after arylsulfatase (80 μg) treatment was centrifuged at $100,000 \times g$ for 1 hr. The pellet was resuspended with 50 mN Tris-HCl (pH 7.5) and incubated with different amounts of sulfatide (2.5, 5, 7.5, 10, 12.5, 20 times the sulfatide content) at 20°C for 30 min. Aliquots (10 μg of protein) were used to determine the enzyme activity. The remaining volume resulting was used to isolate and determine sulfatide and phospholipid content. Enzyme activity is expressed as in Table XVI. The values represent the mean ± SD of five preparations (Zambrano and Rojas, 1986).

Bonting, 1981). Although the transport function of Na$^+$,K$^+$-ATPase and H$^+$,K$^+$-ATPase differ, they have in common some properties and probably much of their mechanism and structure.

5. CONCLUSIONS

Both transport ATPases are typical intrinsic membrane proteins. They depend on a phospholipid bilayer for their activity and transport function. For optimal activity a certain degree of fluidity of the bilayer is required, which is maintained by the presence of unsaturated fatty acids in the phospholipids (Walker and Wheeler, 1975; Harris, 1985). In both enzyme mechanisms the phosphorylated intermediate requires K$^+$ for its hydrolysis. This conformational transition is responsible for inward K$^+$ transport by the Na$^+$,K$^+$-ATPase, and for outward K$^+$ transport by H$^+$,K$^+$-ATPase. These results seem to indicate that sulfatide is the specific lipid requirement of both enzymes and is involved in the affinity sites for K$^+$.

ACKNOWLEDGMENTS. We express our thanks to Dr. Tulio Núñez for his helpful advice and valuable discussions. We are also indebted to Mr. Luis Bondis for his technical assistance. The studies that are reviewed were supported by the Dirección de Investigación y Biblioteca de la Universidad de Chile, under grant No. B-952-8233 and Fondecyt grant No. 1055/85.

REFERENCES

Albers, R. W., 1976, in: *The Enzymes of Biological Membranes*, (A. Martinosi, ed.), Vol. III, pp. 283-301, Plenum Press, New York.
Alvarado, R. H., and Moody, A., 1970, *Am. J. Physiol.* 218:1510-1516.
Baker, P. F., 1972, in: *Metabolic Pathways* (L. E. Hokin, ed.), Vol. VI, pp. 243-268, Academic Press, New York.
Bonting, S. L., and Caravaggio, L. L., 1963, *Arch. Biochem. Biophys.* 101:37-46.
Chen, P. S., Toribara, T. Y., and Warner, H. 1956, *Anal. Chem.* 28:1756-1758.
Dahl, J., and Hokin, L. E., 1974, *Ann. Rev. Biochem.* 43:327-356.
De Pont, J. J. H. M., Van Prooijen-Van Eeden, A., and Bonting, S. L., 1978, *Biochim. Biophys. Acta* 508:464-477.
Drapeau, P., and Blastein, R., 1980, *J. Biol. Chem.* 255:7827-7834.
Edelman, I. S., 1974, in: *Drug and Transport Processes* (B. A. Callingham, ed.), pp. 101-110, MacMillan, London.
Fleischer, B., and Zambrano, F., 1974, *J. Biol. Chem.* 249:5995-6003.
Fleischer, B., Zambrano, F., and Fleischer, S., 1974, *J. Supramolecular Structure* 2:737-750.
Glynn, Z. M., and Karlish, S. J. D., 1975, *Ann. Rev. Phys.* 37:13-55.
González, E., and Zambrano, F., 1983, *Biochim. Biophys. Acta* 728:66-72.
González, M., Morales, M., and Zambrano, F., 1979, *J. Membrane Biol.* 51:347-359.
Hansson, C. G., Karlsson, K.-A., and Samuelsson, B. E., 1978, *J. Biochem.* 83:813-819.

Harris, W. E., 1985, *Biochemistry* **24**:2873-2883.
Jedlicki, A., and Zambrano, F., 1985, *Arch. Biochem. Biophys.* **238**:558-564.
Jørgensen, P. L., 1974, *Biochim. Biophys. Acta* **356**:53-67.
Jørgensen, P. L., 1980, *Physiol. Rev.* **60**:864-908.
Karlsson, K.-A., 1976, in: *Structure of Biological Membrane* (A. Abrahamson and E. Pascher, eds.), pp. 245-274, University of Göteborg.
Karlsson, K.-A., Samuelsson, B. E., and Steen, G. O., 1971, *J. Membrane Biol.* **5**:169-184.
Kean, E. L., 1968, *J. Lipid. Res.* **9**:319-327.
Kyte, J., 1975, *J. Biol. Chem.* **250**:7443-7449.
Lazdunski, M., 1972, in: *Current Topics in Cellular Regulation* (B. L. Horecker and E. R. Stadtman, eds.), Vol. VI, pp. 267-310, Academic Press, New York.
Mehl, E., and Jatzkawitz, H., 1968, *Biochim. Biophys. Acta* **151**:619-627.
Nicolson, G. L., 1976, *Biochim. Biophys. Acta* **457**:57-64.
Nicolson, G. L., and Singer, S. L., 1974, *J. Cell. Biol.* **60**:236-241.
Perrone, J. R., Hackeney, J. F., Dixon, J. F., and Hokin, L. E., 1975, *J. Biol. Chem.* **250**:4178-4184.
Post, R. L., Sen, A. K., and Rosenthal, A. S., 1965, *J. Biol. Chem.* **240**:1437-1445.
Roloefson, B., and Deenen, L. L. M., 1973, *Eur. J. Biochem.* **40**:245-257.
Schuurmans Stekhoven, F., and Bonting, S. L., 1981, *Ann. Rev. Physiol.* **61**:1-76.
Sen, P. C., and Ray, T. F., 1979, *Arch. Biochem. Biophys.* **198**:548-555.
Singer, S. J., 1976, in: *Structure of Biological Membrane* (A. Abrahamson and F. Pascher, eds.), pp. 443-461, University of Göteborg, 1976.
Singer, S. J., and Nicolson, G. L., 1972, *Science* **175**:720-731.
Skou, J. C., 1957, *Biochim. Biophys. Acta* **23**:394-401.
Skou, J. C., 1975, *Quart. Rev. Biophys.* **7**:401-434.
Stein, W. D., Web, W. R., Karlish, S. J. D., and Eilam, Y., 1973, *Proc. Natl. Acad. Sci. USA.* **70**:275-278.
Tsai, K. H., and Lenard, J., 1975, *Nature (London)* **253**:554-557.
Verklaij, A. J., Zwaal, R. F. A., Roelofsen, B., Comfurius, P., Kastelijn, D., and Van Deenen, L. M., 1973, *Biochim. Biophys. Acta* **323**:178-182.
Walker, J. A., and Wheeler, K. P., 1975, *Biochim. Biophys. Acta* **394**:135-144.
Zambrano, F., and Rojas, M., 1986, *Arch. Biochem. Biophys.* **253**:87-93.
Zambrano, F., Fleischer, S., and Fleischer, B., 1975, *Biochim. Biophys. Acta* **380**:357-369.
Zambrano, F., Morales, M., Fuentes, N., and Rojas, M., 1981, *J. Membrane Biol.* **63**:71-75.

Index

Absorption spectrum, 178
Acceptor, 204
Acetylcholine, 157
    binding site, 155
    controlled channel, 157
    receptor, 147, 148
Acid secretion, 230
Acrylamide, 192, 194, 206
Active transport of Ca^{2+}, 177
Adenosine triphosphatase, 212
Amiloride, 216
Anisotroic rotation, 135
Anisotropy, 201
    decay, 139, 201, 202
Annular lipid, 112, 113
Annulus, 112
Antibodies, 154
Antigenic determinants, 154
Apical membrane, 228
Arysulfatase, 217, 219, 228
Atomic coordinates, 15
Autocorrelation functions, 142
Averaged anisotropy, 26
Axial
    diffusion, 51
    symmetry, 141

Axially symmetric shift tensor, 25
Axially symmetric tensor, 24

Bacteriorhodopsin, 95, 97
Bee venom, 98
Bimolecular collision rates, 128
Bimolecular quenching constant, 192
Binary mixture, 90
Binding constant, 109
Bleaching, 126, 127
Bound water, 78
Brownian motion, 197
Bungarotoxin, 151, 157

Ca^{2+}-ATPase, 178, 179
Calcium transport, 177
Calorimeter, 76, 103
Carbamylcholine, 166
Catalytic cycle, 187
cDNA, 151
Cerebroside, 214
Channel gating, 166
Chemical shift
    anisotropy, 22, 23, 42, 141
    Hamiltonian, 23
Cholesterol, 43, 44, 63, 64, 66, 92, 164, 165

Collision frequency, 128, 130
Combustion calorimeter, 103
Conductance, 157
Conformational change, 188
Conformational order parameter, 117
Crosslinking, 159
Crystal powder, 31
Crystallites, 4, 24

Dephosphorylation, 225, 226
Depolarization, 197
Desensitization, 158
Differential scanning
  calorimetry, 71, 72
  calorimeters, 75
Diffusion, 194
  coefficient, 125, 129, 170
Diphenyl hexatriene, 197
Dipole
  dipole interaction, 204
  moment, 190, 191, 196
Donor, 204
Dynamic quenching, 192, 194

Effective volume, 199
Electrocyte, 160, 163
Electron
  density, 10, 11
  spin resonance, 140
Electronic state, 180
Electronic transition, 180
Elementary currents, 158
Emission
  dipoles, 197
  spectra, 178, 275
  wavelengths, 186
Energy
  transduction, 177
  transfer, 182, 197, 206
Enolase, 203, 204
Euler angles, 23
Excimer formation, 128
Exciplex, 183
Excited electronic states, 180
Excited state dipole, 191

Fast anisotropic rotation, 26
Fatty acid composition, 160
Filipin, 164
Five-helix model, 153, 154

Flash photolysis, 137
Flip-flop, 123
Fluorescent nucleotides, 190
Fluorescence, 137
  anisotropy, 201
  depolarization, 198
  emission spectrum, 178
  energy transfer, 204
  lifetime, 182, 194, 198, 199
  photobleaching, 126
    recovery, 126
  polarization, 184
  qenching, 191
  spectroscopy, 178
  yield, 179
Forster distance, 204
Four-helix-model, 153
Free area theories, 170

Gating, 157, 159
Gel phase, 38, 50
Glycolipids, 212
Golgi, 214
Ground state, 180, 183, 194

Harmonic or phase-modulation method, 182
Heat
  capacities, 82, 84, 106
  of dilution, 105
  of incorporation, 112
  of ion binding, 104
  of neutralization, 105
  of reaction, 105, 107
  of transition, 75
Hydrophobic
  domains, 191
  hydration, 83, 84, 106, 111, 112
Hydrostatic pressure, 203
Hyperfine anisotropy, 142
Hyperfine splittings, 141

Immobilized lipid, 168
Intramembranous, 168
Intrinsic fluorescence, 184, 186, 187
Isotropic
  molecular motion, 22
  motion, 25
  rotation, 134
  rotational coefficient, 134
  rotational diffusion, 134
Isotropical chemical shift, 23

L_α phase, 38, 40, 80
L_β phase, 80
Lateral diffusion, 44, 123, 171
Lifetimes, 180
Limiting polarization, 197
Line shapes, 25, 27, 33, 40, 52, 60
Lipid
　annulus, 95
　domains, 160
　metabolism, 163
　mixtures, 41, 59, 84
　monolayer, 165
　packing, 167
　protein interactions, 95
　protein interface, 195
Liquid-crystaline phase, 35
Lyotropic mesomorphism, 77

Macroscopic currents, 158
Magnetic relaxation times, 142
Melittin, 98, 102, 112, 114
Microenvironment, 160
Molecular packing, 16
Motional correlation times, 27, 33
Motionally restricted lipid, 168

Na^+, K^+-ATPase, 212
Neuromuscular junction, 168
N-glycosylation, 153
Nitroxide, 196
　spin labels, 140
Nuclear spin labels, 29
Nucleotide sequences, 152

Orbitals, 180
Order parameter, 47, 48, 113, 137, 139
Orientational potential, 136
Ouabain, 215, 219, 223
　binding, 226

P-nitrophenylphosphate, 223
Patch-clamp method, 158
Phase
　behavior, 77
　diagram, 86, 89, 90, 91
　fluorometry, 183
　L_α, 38, 40, 80
　L_β, 80
Phenylalamine, 178
Phosphoenzyme, 222, 225, 226

Phosphorescence, 181
　anisotropy, 171
Phosphorylation, 186, 188, 191, 223, 225
Photobleaching, 137
Photoreactive probes, 159
Plasma membrane, 214
Plasmalemma, 171
Point mutation, 155
Polarity, 178, 190
Polarization, 137, 196, 198, 200, 203
　anisotropy, 138
　of fluorescence, 197
Polarized
　excitation, 197
　flash photolysis, 139
　light, 197
Polarizer, 198
Postsynaptic membrane, 155
Potassium
　binding, 227
　influx, 222, 379
Powder pattern, 32, 63
Pretransition, 80
Protein/lipid ratio, 168
Proton pump system, 228
Pulse method, 182
Pyrene, 128, 194, 200, 202, 204
　butyryl, 199
　maleimide, 187

Quadrupole
　interaction, 22, 23
　splitting, 30, 31, 141
Quantum yield, 182
Quenching, 184

Raman line, 198
Reaction
　calorimetry, 72, 103
　enthalpy, 111, 114
Receptor particles, 168
Reconstitution, 164, 166
Red blood cells, 222
Reflections, 5, 7
Rigid body order parameter, 117
Ripple phase, 38
Rotation matrix, 23, 24, 26, 38
Rotational
　correlation time, 138, 142, 169, 170, 200, 203
　diffusion, 51, 123, 197, 198, 199

Rotational (*cont.*)
  diffusion coefficient, 133, 135, 144, 198
  diffusion tensor, 135
  mobility, 200
  motion, 54, 134
  relaxation time, 171
  time, 199

Saponin, 165
Sarcoplasmic reticulum, 186, 192, 195, 199,
  205
Saturation transfer ESR, 140
Scattered light, 198
Shift tensor, 24
Single-channel, 159
  conductance, 158
  recording, 158
Singlet, 180
  excited state, 180
Small-angle X-ray diffraction, 151
Sodium
  efflux, 219
  flux, 216, 225
  pump, 219, 222, 227
Solvation lipids, 97
Specific volumes, 83
Spin lattice
  orientation, 180
  relaxation time, 140, 142
Static quenching, 194
Stearic acid spin label, 168
Stopped-flow fluorimetry, 184
Subunit arrangement, 152
Sulfatide, 213, 217, 223, 228
Surface charge density, 108
Synaptic cleft, 153
Synaptic region, 171
Synthetic peptides, 154

T_1 relaxation times, 35
Thermotropic mesomorphism, 77
Time-resolved polarization decay, 202
TNP-ATP, 188
  derivative, 187
Transition enthalpy, 74, 75
Translational diffusion coefficient, 124, 130,
  132
Translational motion, 171
Translocation, 213, 219, 225
Transmembrane
  domains, 154
  helices, 150
  portions, 159
Triplet, 180
  excited state, 180
Tryptophan, 178, 195, 204
Tyrosine, 178

Unit cell, 5, 8

Vant't Hoff
  enthalpy, 74, 97
  equation, 73
  transition enthalpy, 96
Vibrational energy, 191
Vibrational modes, 180
Viscosity, 131, 144, 170, 197, 200

Wigner rotation matrices, 135, 136, 142

X-ray fiber
  diagrams, 4
  diffraction, 4

Zeeman interaction, 30